Die-cutting and Tooling

A guide to the manufacture and use of cutting, embossing and foiling dies, anvils and cylinders

Other Labels & Labeling books:

For the latest list please visit: **www.labelsandlabeling.com**

Die-cutting and Tooling

A guide to the manufacture and use of cutting, embossing and foiling dies, anvils and cylinders

Michael Fairley LCG, FIP3 and FIOM3.

Die-cutting and Tooling

A guide to the manufacture and use of cutting, embossing and foiling dies, anvils and cylinders

First edition published 2016 by:
Tarsus Exhibitions & Publishing Ltd

Printed by CreateSpace, an Amazon.com company.

ISBN 978-1-910507-10-0

Contents

Preface

While there is a considerable amount of information readily available on the label printing processes and on various aspects of printing technology, on origination and pre-press, color management and production workflow, the role and importance of cutting labels to size and shape has generally been less well documented. Yet research has shown that almost half of all pressure-sensitive quality faults resulting in packaging line downtime are related to the label converting and finishing process.

In general, these faults are not caused by the cutting, sheeting, slitting, perforating or specialized tooling that is being supplied and used by the converter, but are more likely to be related to issues such as not ensuring that the die was made for the particular material to be cut, or to poor die handling, incorrect cutting pressure, polluted or worn bearers, deflection of the rotary die or magnetic cylinder, the die bouncing, or even anvil roller or magnetic cylinder wear.

Many factors play a part in the correct ordering, specification and use of cutting dies and other related tooling. Manufactures have made great steps in providing information and guidelines for ordering and using cutting tools of all kinds – whether flat, solid rotary, or flexible dies mounted on magnetic cylinders. But do all converters really understand the need for different cutting angles, or why and how pin or air-eject dies might be required?

Remember too, that cutting dies have to be machined and precisely sharpened to very tight tolerances so that they only cut through the face and adhesive layers of the laminate, while male and female embossing dies need to mate exactly together to form the necessary embossed or debossed surface structure. Foiling dies also need to be very precise in structure and format to provide clean separation of the foil.

There are also applications that require more specialized removable-blade sheeters or perforators, hole punching tools, lineal cutting, or a variety of different kinds of web slitting tools – rotary shear, crush cut and razor.

Key to the performance of much of the label printing and converting process and tooling already mentioned is obviously the manufacture, quality and operation of the many different types of complementary and essential cylinders, anvils, sleeves and rollers that are used in rotary printing and converting, together with all the associated bearers, shafts, gears and side frames.

In short, sophisticated die-cutting and tooling technology in label converting and finishing is one of the critical areas of production which enables quality die-cut and converted labels to be successfully produced and applied in multiple applications, using a wide variety of substrate types, on many different types of presses.

This book sets out to explore and describe modern tooling technology and the materials used, from how the tools are manufactured, their use and applications, how they should be handled and stored, right through to troubleshooting on the production line. A handy Glossary of Terms is also included. Hopefully the book will provide a valuable and convenient reference source for label converters, industry suppliers and label end users that buy and specify labels.

Michael Fairley
Director, Labels & Labelling Consultancy
Founder, Label Academy

About the Label Academy

This book is part of the recommended study material for the Label Academy, a global training and certification program for the label industry. The Label Academy was created by the team behind Labels & Labeling magazine and the Labelexpo series of events.

The Academy consists of a series of self-study modules, combining free access to relevant articles and videos with paid text books (both printed and electronic). Once a student has completed a module, there is an opportunity to take an online test and earn a certificate.

It is expected that a Label Academy qualification will become a standard in the industry – for printers/converters, suppliers, brand owners and designers – and assist in providing a benchmark. In addition to its own training, the Label Academy will aim to become a resource provider to the many existing educational programs in the industry. Accredited training courses will be promoted through the Label Academy website and books will be provided at discounted rates.

The Label Academy concept was pioneered by industry expert Mike Fairley. This was in response to a reduction in the number of dedicated printing colleges and the need to standardize training across the world. The label industry also has its own specific training needs – it has some of the widest range of materials, printing processes and finishing solutions of any printing sector.

We are also working with other training experts and authors to ensure that the Label Academy provides up-to-date and relevant training material for the industry.

The Label Academy is supported by the key trade associations, including FINAT, TLMI and the LMAI.

www.label-academy.com

Label Academy sponsors

Thank you to our founding sponsors, without whom this ambitious project would not have been possible:

Cerm

Cerm designs business automation software solutions to meet the specific demands of flexo and digital narrow web printers. Using the latest technology, our team's focus is on innovation and continuous improvement.

Our automation solutions support each step in the printer's integrated workflow – from estimating to production, shipment and data collection – and provide the feature and functionality printers need to gain efficiency and improve profitability.

Cerm inspires collaboration and helps printers remain competitive in the market and deliver the best products possible. We are proud to sponsor the Label Academy and contribute to the future of the narrow web printing industry.
www.cerm.net

Flint Group Narrow Web

Flint Group Narrow Web has the products, the solutions, and the technical experts to handle any print situation. Providing solutions for food packaging, sustainability, increased bottom line, efficiency, and uptime – delivering the basics needed to run a successful operation, and the expertise to go above and beyond to another level of success.

Our experts provide solutions to your printing problems with the innovative products and services that have made us an industry leader around the world. Wherever you are, we are – available to help you reach your business goals today and into the future.

Continuous improvement is paramount to Flint Group; we are proud to sponsor the Label Academy and the benefits it will bring to the future of our industry.
www.flintgrp.com

Gallus Group

The Gallus Group with its production sites in Switzerland and Germany is a leader in the development, production and sale of narrow-web, reel-fed presses designed for label manufacturers. The machine portfolio is augmented by a broad range of screen printing plates (Gallus Screeny), globally decentralized service operations, and a broad offering of printing accessories and replacement parts. The comprehensive portfolio also includes consulting services provided by label experts in all relevant printing and process engineering tasks. The Gallus Group is a member of the Heidelberg Group and employs around 430 people, of whom 253 are based in Switzerland. The group headquarters is in St.Gallen, Switzerland.
www.gallus-group.com

MPS Systems B.V.

Producing high-quality label printing depends on several factors; one of them is the operator of the press.

As a press machine builder since 1996, MPS Systems B.V. knows how important training and education on subjects like pre-press, label printing and finishing is. For label printers, it is critical that their operators keep up with pre-press and press developments in addition to label trends. Therefore, MPS sponsors the Label Academy, to advance operator's passion for printing, share expertise and help multiply benefits.

The MPS slogans of 'Printers First' and 'Technology with Respect' have always underlined the core philosophy of MPS from press design to operator satisfaction. We develop our presses with a strong focus on user-friendliness and respect for the press operator: Printers First.
www.mps4u.com

HP Indigo

HP Indigo is a global leader in digital printing, with a broad portfolio of digital presses and workflow solutions. Indigo's proprietary Liquid Electrophotography (LEP) technology delivers exceptional print quality for the widest variety of applications including labels, flexible packaging, shrink sleeves and folding cartons. HP Indigo's digital presses match gravure print quality satisfying the most demanding brands.

A division of HP Inc.'s Graphics Solutions Business, Indigo serves customers in more than 122 countries, including many of the top label and packaging converters worldwide.
www.hp.com/go/labelsandpackaging

UPM Raflatac

In a little more than three decades, UPM Raflatac has become one of the world's leading manufacturers of pressure sensitive label materials, developing and leveraging the latest innovations in adhesive technology. Our film and paper label stocks are used for product and information labeling across a wide range of end-uses – from pharmaceuticals and security to food and beverage applications.

We are an engineering driven company with industry-leading products known for their consistent high quality and top performance. We are also known for the high performing supply chain and undisputed leadership in the area of sustainability. UPM Raflatac's dedication to innovation, sustainability and top quality is matched only by our commitment to service excellence. We call it the Raflatouch.
www.upmraflatac.com

About the author

Michael Fairley
Director, Labels & Labelling Consultancy
Founder, Label Academy

Michael Fairley has been writing and speaking about label and packaging materials, technology and applications since the 1970s, both as the founder of Labels & Labeling and other print industry magazine titles and as an international consultant writing or contributing to label industry market and technology research reports for the likes of Frost & Sullivan, Economist Intelligence Unit, Pira, InfoTrends and Labels & Labelling Consultancy.

He is the author of the Encylopedia of Label Technology, co-author of the Encylopedia of Brand Protection, a contributing author to the Encylopedia of Packaging Technology and a contributing author to the Encylopedia of Occupational Health and Safety. He also provided significant input to the Academic American Encylopedia.

He now works as a consultant to Tarsus Exhibitions & Publishing – which organizes the Labelexpo shows, Label Summits and publishes Labels & Labeling magazine – as well as regularly speaking at industry conferences and seminars.

He is a Fellow of the Institute of Packaging / Packaging Society, Fellow of IP3 (formerly the Institute of Printing), a Freeman of the Worshipful Company of Stationers, an Honorary Life Member of FINAT and a Licentiate of the City & Guilds of London Institute. He was awarded the R. Stanton Avery Lifetime Achievement Award in 2009.

Acknowledgements

Much of the early information about die-cutting and tooling technology was written by Ron Spring during the 1980s. As the co-founder of Labels & Labeling magazine and also the managing director of Gerhardt UK at that time, he was very much involved in educating the industry and promoting the fast evolving developments in solid rotary dies, spark erosion technology and the introduction of flexible dies. Many articles were published in Labels & Labeling at that time, as well as a wide range of technical literature.

Much of that early written information has now been completely updated or re-written to reflect current industry technology, standards and practices using the substantial Labels & Labeling archive of features, articles and news items. Thanks are due to all those that have written these features, contributed articles and PR stories, and placed related advertisements. Their input has been invaluable.

Additional sources of reference have included conference and seminar presentations, industry supplier websites, Labelexpo and other industry shows, as well as personal contact and discussions with industry leaders, all of which have been much appreciated.

Grateful acknowledgement is made to ABG International, Electro Optic, Gallus Group, Harper, Kocher + Beck, MPS, RotoMetrics, SEI, Schober, UEI Falcontec, and Wink for providing illustrations used in the book, as well as to Greg Smye-Rumbsy for his expert drawing of the many technical diagrams.

Finally, thanks are especially due to UEI Falcontec for kindly reviewing the chapter on embossing and providing feedback and additional material, and to RotoMetrics who extensively read through and reviewed all the die-cutting and related chapters, added further material and illustrations, and ensuring information on performance, testing, handling and storage was as current and accurate as possible.

Many thanks to everyone in the industry that has supported the writing and production of this die-cutting and tooling book. Without their support, encouragement and help its publication would not have been possible.

Chapter 1

The label printing and converting process

The production of pressure-sensitive labels involves a quite complex process of printing with up to ten or more printing units, sometimes using two or more different printing processes. Ancillary operations typically follow as needed to die-cut the labels to shape, remove the waste material, trim the edges of the web and slit it into single roll widths – or sometimes into single sheets – prior to a final inspection, re-winding and then packing before shipping to the customer.

Depending on the specific label application, the types of labels to be produced and their added-value requirements or complexity, it may also be necessary to undertake a variety of other types of finishing or embellishing operations, such as varnishing, embossing, hot or cold foiling, perforating, scoring, sheeting or hole punching.

Many of the finishing operations mentioned above may be carried out in different ways. For example, die-cutting may take place using flatbed cutting dies, solid rotary dies or flexible wrap-around dies – even laser cutting today. Perforating may also be undertaken from a flat die or a rotary die. The same applies to embossing or hot foiling where there may also be segmented dies.

The challenges involved in manufacturing and using cutting, perforating, foiling, embossing or punching dies is that labels – and unsupported film – may frequently be required to be produced using different types of paper, film, foil, metallic, etc., substrates, with different types of liners to cut to, as well as different types of adhesive. Some substrates and adhesives may well be more abrasive or challenging to cut than others, perhaps requiring different cutting angles or bevels.

Understandably therefore, the manufacture of the different types of dies and tooling is a highly specialized and quite complex process, or sequence of processes, using hardened metals and coatings or surface treatments to give different types of quality, performance and cutting profiles.

Many or most of the different types of cutting, punching or perforating dies need an accurate solid surface to cut against, – called an anvil or base roll – or additionally, in the case of flexible dies, a magnetic cylinder on which they are held firmly in position. In some printing processes, flexible printing plates may be used, again held in place on a magnetic print cylinder, or perhaps they are in the form of a sleeve fitted onto a sleeve cylinder. Gear wheels and bearers may need to form part of these various cylinder constructions.

The flexographic printing process has an additional requirement for an anilox roll that carries the fluid ink from the ink tray or pan to the printing cylinder. Again, precision made, anilox rolls are quite complex in their manufacture and requirements and have a significant impact on the quality of printing that can be produced.

The manufacture and production of many of these

ancillary manufactured products used on a pressure-sensitive roll-label and some other narrow-web presses is commonly known as tooling and the items themselves as tools. Put together and depending on the specific supplier or sector, tooling can generally be categorized under one or more of the headings shown in the following flow chart (Figure.1.1):

technology.

Historically, flat die-cutting tools were the dominant technology used on the intermittent feed and semi-rotary presses of the 1970s and 1980s. As rotary printing started to become the major label printing process so the industry moved to solid rotary dies. More recently, flexible dies have become the die-

TOOLING FOR NARROW-WEB PRESSES

Die-cutting tools	Embossing tools	Hot foiling tools	Other tools	Magnetic cylinders	Printing tools
Flexible dies	Flatbed dies	Hot foiling rotary dies	Perforating and punching tools	Magnetic cylinders for cutting dies	Print cylinders
Solid rotary dies	Solid roary dies	Hot foiling flatbed dies	Removable blade sheeters	Magnetic cylinders for hot foiling dies	Anilox rolls
Flat bed dies	Flexible die plates	Flexible hot stamping dies and sleeves	Base rolls (anvils)		Impression cylinders
			Lineal scoring & slitting tools		Meter rolls

Figure 1.1 - Flow chart showing the main types of tooling available for narrow- to mid-web label and related presses

The nature, role and function of each of these types of tooling is outlined below to put them into context within the book and each will then be described in more detail in the chapters that follow.

THE NATURE, ROLE AND FUNCTION OF TOOLING

Before moving on to understand the materials, technology and production of the different types of tooling it would be of value to a know a little more about each type of tool and to be able to recognize what they look like and how, when and where they are used in label printing and converting operations, starting with the various types of label die-cutting tools.

Die-cutting tools are used in the process of cutting a label to size and shape with a die. Most self-adhesive labels have to be die-cut to shape as part of their manufacturing and finishing procedure. Depending on the type of label and the printing and/or die-cutting requirement, this operation may be performed using flexible dies, solid rotary dies or flat dies or, most recently, with digital laser die-cutting

cutting process of choice, overtaking solid tools.

Today, flexible dies are said to account for more than 80 percent of all label dies used in Europe and around 70 percent in North America, largely due to their much easier transportation, storage and handling, as well as not insignificant cost savings. An increasing trend towards shorter run lengths has also influenced the trend towards flexible dies.

Figure 1.2 - Magnetic cylinder and flexible die. Source: RotoMetrics

Flexible dies are thin plates made from spring steel that have the die-cutting configuration etched over the surface and milled to the right dimensions. They are then mounted for use by wrapping the thin steel around a magnetic cylinder.

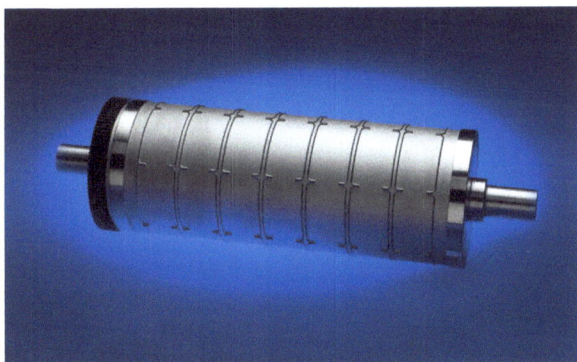

Figure 1.3 - Solid milled rotary die. Source: RotoMetrics

Solid milled rotary dies are engraved from a cylinder of steel so as to leave the cutting edge standing proud around the cylinder circumference.

Each of the types of flexible or rotary die require some form of final finishing following on from the machining or etching, which is undertaken using computer-guided equipment and which sets the seal on the final quality on the die.

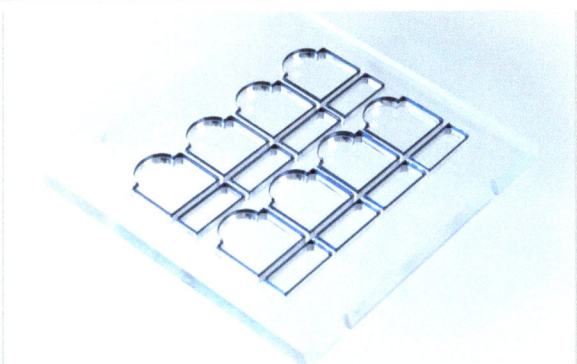

Figure 1.4 - Flat rule die. Source: Wink

Flat dies are most commonly produced by bending lengths of accurately fashioned steel rule, which has been finished to a cutting bevel along one or both edges. This rule is around 0.4mm (0.0157") in thickness and nominally 12mm (0.472") in height.

To form a cutter, the rule, once bent to shape using a special bending tool, is placed in a base into which the shape or shapes of the label(s) has been cut. In this way the rule is supported during use in the die-cutting and is able to retain a high degree of accuracy.

A magnetic base and a flat flexible cutting die may also be used for some die-cutting applications.

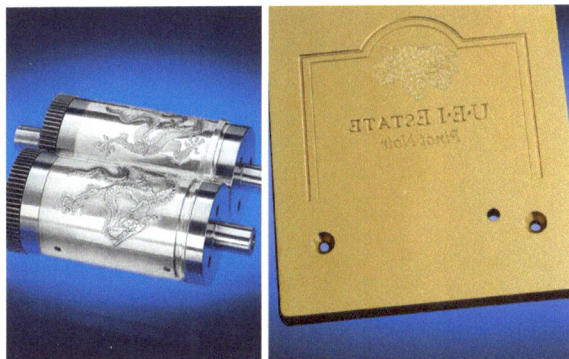

Figure 1.5 - UEI Falcontec Unisphere aluminium rotary embossing dies (left) and flat brass embossing die (right)

Embossing tools. Embossing tool dies are used to shape/set the surface of a substrate to create a raised (embossed) or recessed (debossed) design. Embossing can be achieved with a matched male die and female counterforce (flat or rotary). The female counterforce has the required image incised into the surface; the male die has a matched raised image. An offset is applied to one or both images to enable the dies to accommodate the thickness of the material being embossed.

During the embossing process, the two dies are pressed together through a substrate to create a raised image.

In the process of debossing - which is the reverse of embossing - the positions of the relief die and the counter die are reversed.

3

Male/female embossing dies have their widest application in rotary embossing, although other approaches may be used for certain applications.

For less detailed images, more economic single male or female embossing dies used in conjunction with a hard rubber coated anvil may well achieve the desired result.

Embossing dies are produced to run in both dedicated embossing units or in an available cutter station.

Figure 1.6 - Solid rotary hot foiling cylinder. Source: Kocher+Beck

Foiling tools. Hot foiling or hot stamping is a dry printing process which uses very thin metalized foil in a variety of metallic colours – such as gold, silver, red or blue – rather than inks from which to print. The hot-foil printing process is achieved by transferring the colored metallic pigment coating from a ribbon of plastic material known as the 'carrier' onto the surface of the label material to be printed using a solid or flexible printing die or plate which bears the image to be hot-foiled.

The transfer is achieved through the application of heat, pressure, and the length of time the heated coating area is in contact with the substrate – known as the dwell time. The balance and control of these elements is critical and must be individually calculated for the surface to be printed, and the type of ribbon or foil being used.

The printing plate used for hot foil blocking needs to be of a hard material and have a raised image similar to that used by the letterpress process. The fact that image transfer relies upon both heat and pressure restricts plate materials to either a very hard thermoformed plastic plate for very short runs or plates produced from brass, copper, magnesium, steel, or zinc, for the longer runs.

Figure 1.7 - Econofoil/Unisleeve hot-foiling flexible dies designed to run on low cost aluminium mandrels. Source: UEI Falcontec

Hot-foil blocking/stamping is used on both short and long runs today. Further, traditionally hot foil stamping was used on rotary presses with cylinders; however, today flexible dies are used on rotary presses and flat dies are used on other narrow web presses.

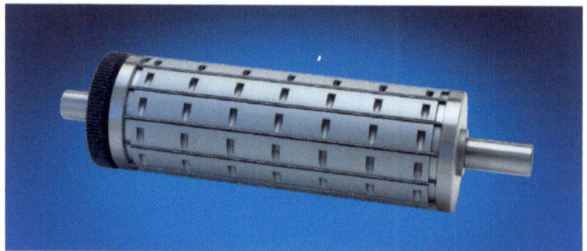

Figure 1.8 - A (cross) perforation cylinder/sheeter. Source: RotoMetrics

Perforating and sheeting tools. In rotary form, perforation tools are cylinders that contain removable or floating blades that can be used to produce perforated or cut lines (sheeting or scoring). They are designed for cutting or perforating across a web to produce items such as A4 sheets, or to provide perforation lines that are punched into a label surface for, say, fan-folding.

The perforating and cutting blades themselves are able to be positioned equally stepped, or freely positioned around the circumference of the cylinder to provide for production flexibility. The blades are held in place by clamping or counter bars and set screws.

Figure 1.9 - Hardened steel anvil roller. Source: Kocher+Beck

Anvil rollers. An anvil roller is a hardened steel roller upon which the bearers of a rotary die, magnetic cylinder or perforation cylinder run.

Normally, this cylinder is placed in the bottom position of a die station. However, for some jobs it is necessary to place the anvil roller in the top and the cutting cylinder in the bottom position. In case of a support roller the anvil roller would be in the middle position.

Anvil cylinders/rollers (Figure 1.9) are characterized by their exemplary hardness and run-out accuracy, whether in use as a standard diameter or a plus or minus cylinder so as to compensate for a wide range of backing papers.

Straight anvils are used for most standard daily operations, and stepped anvils are used to extend the life of a die when most needed and to provide the flexibility to run dies on liners other than those for which they were made, or to compensate for worn cutting tools.

Hole punching tools. A hole punching (pin feed) tool was historically used for making EDP (electronic data processing) holes but other shapes are also possible. The holes, which were used to guide material in finishing lines through dot matrix, continuous laser or thermal printers, are cut by using either a male/female system or using a shaft with movable EDP rings. The latter is used in anvil position to cut up to the face material, with the waste being removed by the waste matrix system when stripping the face.

Figure 1.10 - Hole punching shaft with movable EDP rings. Source: RotoMetrics

The punches are held in position with a set of screws, tightened into a groove. Rings can be adjusted across the web.

There are also microhole punching tools which are designed to create tear-off holes in postage stamps. The punched waste is removed through the hollow die cylinder with working widths up to 500 mm. The modules are connected mechanically or electronically to the converting machine.

Figure 1.11 - A RotoMetrics printing cylinder with bearers and gear wheel

Printing cylinders. Standard printing cylinders form the basis of every label printing machine. These cylinders, together with hot foiling and embossing cylinders, are all manufactured with the greatest care and finite precision in order to guarantee optimum fit and run-out accuracy.

In the flexographic, letterpress and litho processes the printing plates are located on the print cylinders. Each cylinder needs to have accurate and even contact with the inking rollers and the surface of the substrate, or in the case of the litho process, the offset blanket.

Printing cylinders (Figure 1.11) used in the roll-label industry include plate cylinders, blanket cylinders and impression cylinders and these are made from solid aluminium or steel, or produced as a tube with end rings fitted, and with spur or helical gears.

There are also a number of suppliers offering unique coatings applied directly onto existing or newly manufactured printing cylinders. These coatings can provide additional surface protection thereby potentially increasing the life of the cylinder.

Figure 1.12 - Magnetic cylinder and flexible die. Source: Kocher+Beck

Magnetic cylinders and bases. Magnetic cylinders used with flexible dies provide an economic alternative to standard rotary die cutting tools. They are manufactured on CNC machines from high tensile and high alloyed stainless tool steel with fully hardened bearers. Hard ferrite or ceramic and rare earth permanent magnets, hardened bearings seats

as well as bearing necks (journals) with fully hardened centering sleeves are usually standard. Higher-strength magnet configurations are available based on the application.

Magnetic cylinders are available for a full range of label presses and converting machinery, allowing converters to use flexible dies in many different applications.

Figure 1.13 - A RotoMetrics magnetic flat base (left) and a close up of a section from a Kocher + Beck magnetic base (right)

Where flexible dies are used in flat die-cutting systems, then flat magnetic bases can be provided by a number of suppliers. Magnetic bases are used for a variety of different applications, particularly those involving intricate patterns or shapes and can offer a more accurate total die height when compared to most types of steel rule dies.

Flatbed magnetic base systems can also be used for hot stamping and embossing but for foil stamping, they would need to be a heated base.

Figure 1.14 - RotoMetrics RD razor slitter

Slitting of label webs. Once printed and die-cut using flat, rotary or flexible cutting dies, the web of labels (2, 3, 4 or more) printed labels across the web needs to be subsequently converted into individual label widths so that each individual label can be removed from the backing liner in the subsequent label application process. This is because most label presses, depending on label size, are producing more than one label across the web width, which means that the printed and die-cut web will need slitting lengthwise at some stage into individual web widths for rewinding into the final applicator-sized reels.

Figure 1.15 - An ABG International scissor knife slitting unit in operation

Some webs may also need an unwanted edge on either side of the printed web to be removed, commonly known as edge trim. Slitting operations, either in-line on the press or off-line on a slitter rewinder, are commonly undertaken in a slitting unit which utilizes cutting heads that can be of a crush cut, razor (shown in Figure 1.14) or rotary scissor construction. Figure 1.15. shows a typical scissor knife slitting unit found in operation on a roll-label press or on a finishing line.

Anilox rolls. Anilox tools or rolls consist of an engraved metal or ceramic-coated roll used to meter ink to the raised (image) areas on the relief printing plate used in the flexographic inking system. Each type of flexographic press uses an anilox roll, the surface of which is engraved at one of three angles with a pattern of tiny cells of fixed size and depth that transfer the ink to the plate. The cells are so small

that they can only be seen under magnification. The size and number of these cells determine how much ink will be delivered to the image areas of the plate, and ultimately to the substrate. An anilox roll today is either copper-engraved and then chrome-plated, or ceramic-coated steel with a laser-engraved cell surface.

Anilox tools are carefully selected for specific types of printing, substrates, and customer requirements. The printer may well perform test runs to determine the ideal anilox for producing the desired ink distribution for halftones, spot color and solids.

There is also some use of anilox sleeves. These are not new. They have been under development, testing, trials and use for a number of years, with many manufacturers of 'gearless' presses particularly looking at the benefits of sleeve technology, such as ease of register, overall quality, low weight, maintenance, etc., as well as lower shipping costs and storage capabilities.

Figure 1.16 - An anilox sleeve. Source: Harper

Although early attempts at anilox sleeve manufacture were somewhat hit-and-miss due to a variety of reasons there have subsequently been dramatic improvements in the material, construction and stability of anilox sleeves.

In use, sleeves need to be expanded to fit securely on the press mandrels. This may be undertaken with a mechanical mandrel that expands by hydraulic action, or through the use of a press mandrel with an inner compressible layer, or bladder, that is activated by air.

Mandrels used with anilox sleeves are critical to

the success of the process, with diameter, circularity and concentricity all important.

CONFIGURING TOOLING IN A LABEL PRESS

Modular presses are the most common form of label press configuration used by the label converter today. Such presses start with an unwind unit and web infeed, incorporate web tension control and are then followed by the various printing units and processes that make up the converter's press specification. Almost any printing process may be used, either on its own as a dedicated 'single' process machine or in the form of a multi-process combination press that is capable of total process interchangeability.

Within the printing section of the label press the 'basic' tooling products that can be seen are the print cylinders, anilox rolls in flexo presses. Where anilox rolls are used these are manufactured by specialized anilox roll suppliers and are not included within this particular handbook.

When it comes to the finishing section of the label press this is where the main types of tooling described in this handbook are to be found. Most common of the 'finishing, embellishing and other converting processes' which are compatible with the pressure-sensitive label printing processes and found to some degree on almost all label printing and converting lines are:

- Die-cutting
- Hot foil stamping
- Cold foiling
- Embossing
- Perforating
- Hole punching
- Sheeting
- Slitting

Die-cutting is carried out in a cutting station or unit, which may be flatbed, semi-rotary or rotary depending on the particular type of label press, and this is followed by the removal of the matrix waste. A diagram showing a rotary die cutting unit and matrix waste removal can be seen in the following illustration (Figure 1.17).

The die-cutting tools will cut against an anvil roller (crush cut) and the two cylinders are typically stacked on top of each other in a conventional cutting unit. Typically, there is a support roller below the anvil roller.

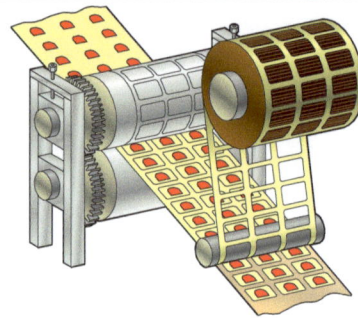

Figure 1.17 - A typical rotary die-cutting and matrix rewinding unit

The repeat size of the cutting cylinder will vary from job to job. A cutting unit is generally constructed so that the cutting cylinders are relatively easy to replace.

Put together, it can be readily seen that a variety of highly accurate and specialized tooling is needed to produce pressure-sensitive labels on a roll label press.

A typical press construction and location of some of the more commonly used tooling is shown in the Figure 1.18. More information on the structure of a printing press, the printing processes, and print finishing can be found in the 'Conventional Printing Processes' book.

This book is primarily concerned with tooling that is used at the finishing end of a label press rather than the manufacture and use of print cylinders and anilox rolls that are used in the analog or digital press section (although these are both discussed in Chapter 7).

The nature and use of cutting, embossing and foiling dies, speciality dies and tooling, slitting wheels or blades, as well as magnetic cylinders and anvil rollers, are all set out in the following pages, together with guidelines for optimizing the cutting processes and for handling, storage and health and safety in the use of tooling.

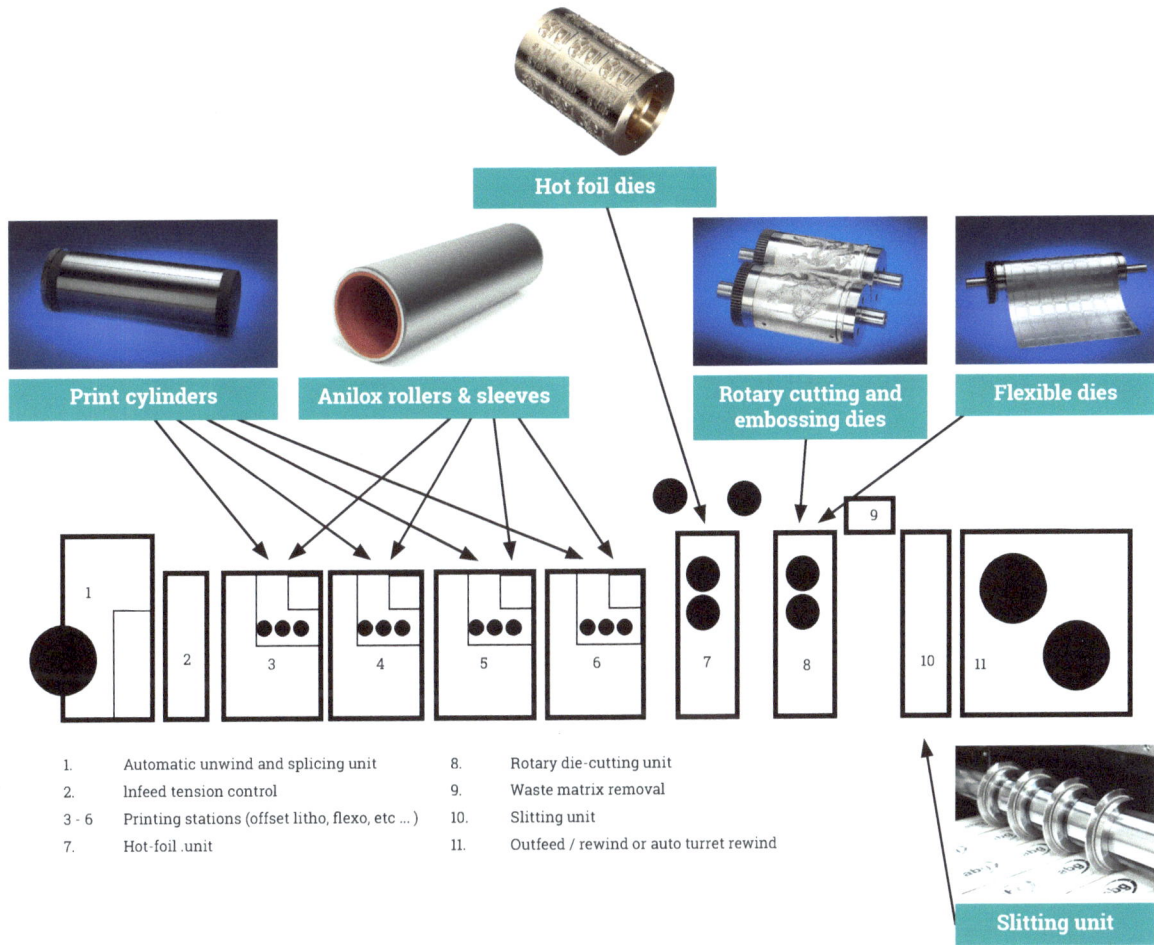

Hot foil dies

Print cylinders

Anilox rollers & sleeves

Rotary cutting and embossing dies

Flexible dies

1	2	3	4	5	6	7	8	9	10	11

1. Automatic unwind and splicing unit
2. Infeed tension control
3 - 6   Printing stations (offset litho, flexo, etc ...)
7. Hot-foil .unit
8. Rotary die-cutting unit
9. Waste matrix removal
10. Slitting unit
11. Outfeed / rewind or auto turret rewind

Slitting unit

Figure 1.18 - Shows in diagrammatic form the construction of a flexographic roll-label press and where some of the most common forms of tooling would be found

Chapter 2

Die-cutting of label webs to shape and size

Pressure-sensitive labels are printed and converted on a roll-label press that takes a reel of pressure-sensitive material, unwinds it and then prints from multiple print heads. The printed web may also be foiled, embossed, varnished or otherwise processed or converted before the labels are cut to shape and size, the reel slit into single label webs and re-wound ready for sending to where the labels will eventually be applied.

Depending on the type of label and the printing and/or die-cutting requirement, the mechanical die-cutting operation may be performed using flexible, solid rotary or flat dies made from steel or, most recently, using digital laser die-cutting technology.

The process of cutting and separation of the processed material using cutting tools is known as die-cutting (shearing) and is probably the most critical in terms of the final end-use application of the label. This shearing or die-cutting is known as a non-

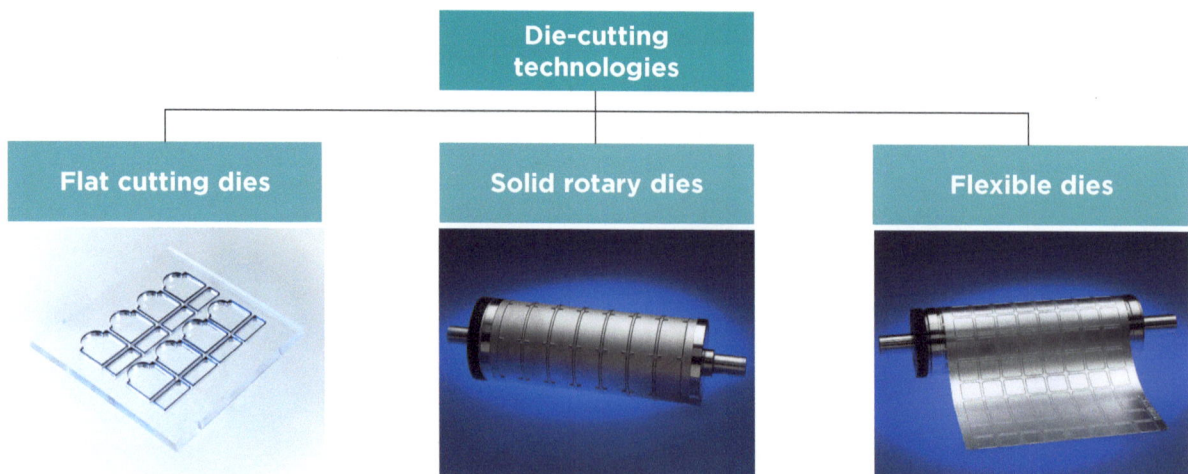

Figure 2.1 - Types of cutting dies for pressure-sensitive labels

chipping process (no material is removed during the separation).

Historically, flat steel die-cutting tools were the dominant technology used on the early intermittent feed and semi-rotary presses of the 1970s and 1980s. As rotary printing started to become the major label printing process so the industry moved to solid rotary dies. More recently, flexible dies have become the die-cutting process of choice, overtaking solid tools for many applications.

In operation, the cutting die must cut cleanly through the label face material and the adhesive layer, but must not cut into or through the silicone coated backing release liner. Die-cutting is therefore an operation that requires precision and fine control of the cutter-manufacturing process. This would include the choice of cutter materials and their properties as well as the cutter set up in the printing and converting machine. In turn, then, a constant thickness is required for the laminate and the release liner backing material. Figure 2.2 illustrates how the matrix waste needs to be peeled away cleanly after the die-cutting process has taken place.

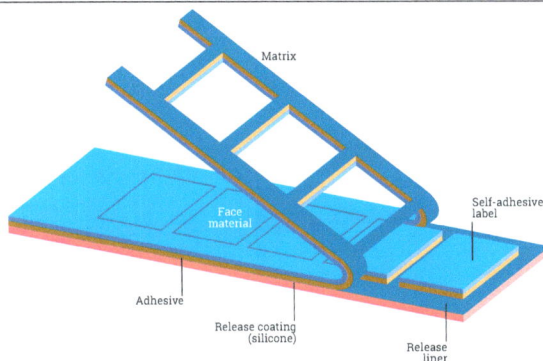

Figure 2.2 - Removal of matrix waste cleanly after die-cutting

Essentially, the cutting die – whether flat, solid rotary or flexible – has to be tooled to the thickness of the backing liner and, if this varies and the die cuts into the liner, as in Figure 2.3 (known as die strike), then the die will have to be re-set or the material evaluated for required settings. If the die-cutting is too

heavy, or too light, and the adhesive has not been cut through cleanly, then labels will not dispense correctly and may not even separate at the beak or stripper plate on the label applicator, instead remaining on the backing liner and staying on the web. Thickness variations from place to place in the backing liner will also cause problems.

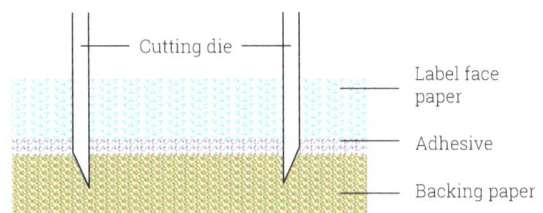

Figure 2.3 - Die-cutting into the backing liner

If the adhesive has not been cut through cleanly or has bled, or there are blobs on the surface, then further problems may arise due to the applicator rollers gumming up.

In this chapter we will be examining the properties of die-cutter tooling materials and explaining the die-cutting process, either performed using flatbed dies, rotary solid dies or with flexible dies.

DIE-CUTTER TOOLING MATERIALS

Die-cutting tools, whether flatbed, semi-rotary or rotary, are prepared using tooling steels that require specific properties, particularly the following:

- Toughness
- Wear resistance
- Heat resistance
- Hardness

The first three of these properties are inherent to tool steel. Hardness is another property in tool steel that is developed through a heat treating process. So let's examine these requirements before looking at the different types of cutting dies and their manufacture and performance.

Toughness is the ability of steels to resist cracking, chipping or breakage. Toughness is usually measured in foot-pounds of energy required to break the steel. Steel in its simplest form is very tough. As alloying content is added, steel loses some of its toughness. Low alloy grades of steel will have better toughness than higher alloy grades.

Wear resistance. A tool is in constant movement, either against another tool or work material. This constant movement can wear away the surface of the cutting tool. The tool will therefore need to have some amount of wear resistance. Wear resistance is the steel's ability to resist erosion. Wear resistance in tool steel is achieved by the presences of carbides. Chromium, molybdenum, tungsten and vanadium are the four carbide-forming elements commonly found in tool steels.

The various carbides are listed in the order of the hardness of the carbides that they form. So a tool steel with vanadium carbides should have more wear resistance than a tool steel with the same amount of chromium carbides. Many tool steel grades will have more than one type of carbide to balance with the other properties. Wear resistance testing is non-standardized, but the grades are always compared in relation to each other.

Heat resistance is the third inherent property of tool steel. It is the steel's ability to resist softening when exposed to heat while in operation. This is a very important property in high-speed steel tools. Excessive heat in a turning or milling operation can lead to the softening of the tool. This softening could allow a tool to dull or chip causing premature failure. With the addition of tungsten and/or cobalt in high-speed steel, heat resistance is improved.

Hardness is a very important property of tool steel that is developed during the heat treating process. It is not one of the inherent properties of tool steel. Hardness is developed through the addition of carbon. A minimum of 4 per cent carbon is needed in order to achieve the hardness necessary for tool steel. Hardness is the steel's ability to resist compression, indentation or deformation. Without hardness the tool would collapse. Hardness is directly related to compressive strength. Hardness is typically measured on a Rockwell C-scale. Tool steel at Rc 62 will have higher compressive strength than tool steel at Rc 55, regardless of the grade of steel chosen.

Subsequent hardening of tooling steels can be carried out using case hardening, induction hardening, vacuum hardening or laser hardening, briefly explained as follows:

- Case hardening or surface hardening is the process of hardening the surface of a metal object while allowing the metal underneath to remain soft, thus forming a thin layer of harder metal (called 'the case') at the surface. Case hardening is usually done after the part has been formed into its final shape.
- Induction hardening. This is a form of heat treatment in which a metal part is heated by induction heating and then quenched. The quenched metal undergoes a transformation, increasing the hardness and brittleness of the part. Induction hardening is used to selectively harden areas of a part or assembly without affecting the properties of the part as a whole.
- Vacuum hardening. Vacuum hardening of solid tooling is mostly done in a so-called vacuum furnace which is a type of furnace that can heat materials, typically metals, to very high temperatures and can carry out processes such as heat treatment with high consistency and low contamination.
- Laser hardening. As an alternative to the more common surface treatments, some suppliers of flexible dies also offer laser hardened dies. Laser hardening is not offering any reduction in the friction coefficient, but does offer a partial increase in hardness at the tip of the cutting edge. The increase in hardness depends on the carbon content in the steel and not on the energy put in by the laser.

Tool steel selection. Selection of a tooling steel is a balance between the four properties already described above. No single tool steel will provide the best of all of these properties. Other factors also play a role in tool steel selection. Every one of the tool steel's used must be machine-able in order to form it into a useful tool. Machinability is usually expressed as a percentage of the most basic tool steel, which would be valued at 100 per cent machine-able. Many tools will also require some degree of grind-ability.

Tools are usually manufactured to very close tolerances and grinding is performed to achieve these close tolerances.

Tools may also require some degree of polish-ability to allow the manufactured part to release from the tool. Fine-grained or powdered metal tool steels allow the highest degree of polish-ability. Some other important properties of tool steels include distortion and safety in heat treating, austenizing temperature range, price and availability.

FLAT DIE-CUTTING PROCESS AND TOOLS

In the earlier days of pressure-sensitive label production, all cutting units were flatbed. A flatbed cutting unit is characterized by the fact that the converted material is not moving relatively to the cutting tool. The cutting tool is punching in to the material while the material is stationary during the cutting operation. After the cutting sequence has been completed, the material is moved forward to the next position, where the cutting sequence will be repeated and so on. Flatbed cutting dies are used on all label printing machines with flatbed units, including semi-rotary label presses (Figure 2.4).

In one sequence, the entire cutting rules of the cutting tool are being punched in to the material at the same time. This will require a relatively high cutting force because of the amount of cutting taking place in one operation at the same time. In a flatbed converting unit, the cutting depth has to be adjusted to the actual thickness of the face material and adhesive.

As can be seen in the diagram, the configuration of a flatbed die-cutting operation consists of three main elements:

- **The in-feed:** Which takes control of the web and registers it to the cutting tools
- **The flatbed platen section:** This is the heart of the unit, bringing the cutting die and cutting plate together under pressure. Held between them, the substrate may be flatbed die-cut, creased, or even perforated depending on the particular application and substrate
- **The stripping section:** Matrix waste, and sometimes side and rear trim, is removed using a matrix re-wind spool or roller.

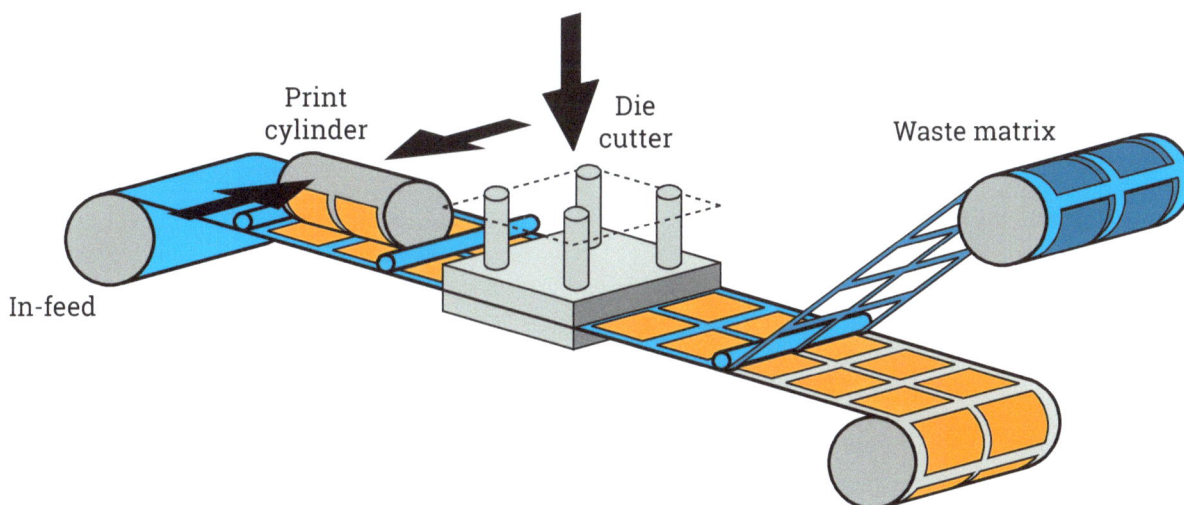

Figure 2.4 - Semi rotary stop feed label press

If we look at pros and cons for this type of flatbed die cutting unit we can see the following key advantages and disadvantages:

Advantages

- The cutting tools used for this type of cutting are relatively simple and therefore also relatively inexpensive.
- Because a flatbed cutting unit has the ability to adjust the cutting depth, the same cutting tool can be used for different material thicknesses and can both kiss cut and cut through the material.
- Is good for short run work.
- Offers easy access for the operator.

Disadvantages

- Compared to rotary cutting, flatbed converting is a slower process and the initial adjustment of the cutting tool is more time consuming than in a rotary converting process.
- They are not as robust as solid dies.
- Mainly for productivity reasons, flatbed die-cutting technology is less widespread today in the pressure sensitive industry.

Cutting tools for flatbed die-cutting units typically consist of steel rules which are held in place in grooves cut to shape in a wooden or plastic base material. Nowadays, flexible dies are also used for flatbed die-cutting in conjunction with a precision-made magnetic base plate that holds the flexible die in position.

Because of the force required for flatbed die-cutting – with the entire cutting operation undertaken simultaneously – the flatbed die-cutting unit is usually a heavy and solid piece of equipment with very good stability for the cutting process.

Steel rules used for flatbed die-cutting of labels are usually positioned to the desired label shape(s) by using a jigsaw, vertical milling machine or a laser beam to cut grooves in a plywood or plastic base. Cutting rule is then cut and bent to fit the cut out label shape and inserted into the base. Wherever practical, for facilitating removal of the waste, rule joints should be positioned so that they run with the web direction and not across the web.

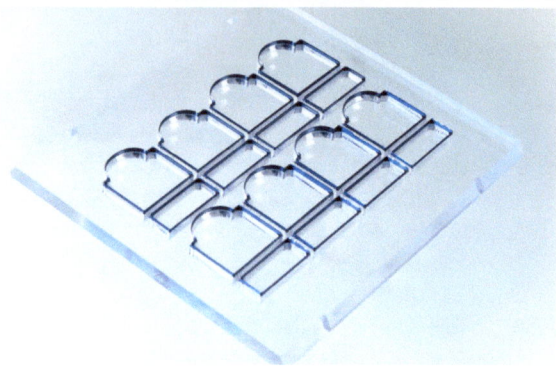

Figure 2.5 - Flat rule die. Source: Wink

Ejection rubber is fitted to ensure the die-cut web separates quickly and cleanly from the cutter. The amount of rubber required will vary according to the complexity of the label shape. Rubber strips are placed on either side of the steel rule to act as the stripper plate; the rubber compresses on the down-stroke and on the up-stroke it pushes the workpiece out of the die. The rubber used varies in thickness, structure and hardness (Shore or Durometer). It is important to use ejection rubber that is slightly higher than the actual height of the cutting line. This will cause compression (sufficient energy and ejection force) on the rubber thus ensuring complete ejection of the cut out part. Base plate height is commonly between 5 – 10 mm.

Trouble free die cutting of labels requires cutting rules with specific properties. The main requirements are extreme dimensional accuracy, edge straightness and minimal tolerances, sharp cutting edges and bevels (cutting angles) suitable for cutting all types of pressure-sensitive materials, long life, seamless rule joints and easy processing on auto-easy benders. The cutting angles of the rules are adapted to the particular pressure-sensitive or other material to be cut and can vary from 50° up to 110°.

In reference to the cutting rule die blades, the bevel(s) of the cutting edge determine how a blade penetrates and cuts the specific material being die cut – paper, plastic, foil, card, soft, hard, etc. A center bevel for example, may be used for general purpose or cardboard cutting, a side bevel for cutting more rigid materials, a centre face double bevel for the converting of thick or harder materials. See Figure 2.6.

The bevel on a die blade is identified by the number of degrees on the bevel or as an inside bevel or outside bevel. Some steel rule blades have a double bevel for specific cutting applications.

Cutting rules are also available in various grades

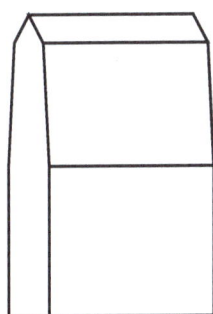

CENTRE FACE DOUBLE BEVEL

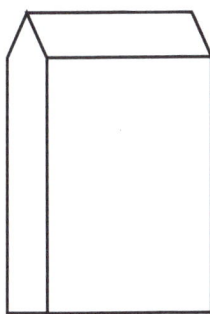

CENTER BEVEL

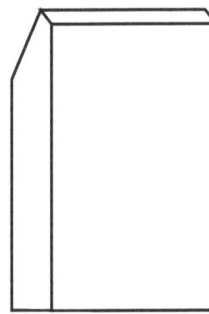

SIDE BEVEL

Figure 2.6 - Examples of different types of steel rule flat die cutting bevels

Quite simply, the bevel on a die blade depends on its intended use. The bevel is generally pre-shaped in the blade and can have a short or long bevel, a center bevel, a side bevel or multiple bevels depending on what the blade is to be used for.

of performance and hardness (described earlier) – including through hardening to give a uniform high hardness on the cutting edge and in the body, induction hardening of the cutting edge and special coatings, all of which can extend the life span of steel rules up to five times, in turn making them particularly economic for multiple repeat jobs and very long life. Label cutting also requires excellent bendability, reduced cutting force and long tool life.

ROTARY DIE-CUTTING TOOLS

Rotary die-cutting uses a solid curved cylindrical or wraparound die in a rotary die-cutting unit, with the cutting taking place against an anvil roller. See Figure 2.8. While the global focus of label die-cutting today has largely moved, or is moving, to rotary die-cutting using flexible dies, solid steel rotary cutting tools still have a significant presence in the industry, and probably will still have a role to play for many years to come. Some substrates such as thicker papers or films are just not suited to cutting with flexible dies; they are too thick. Indeed, the type of material being cut makes a huge difference when it comes to die selection.

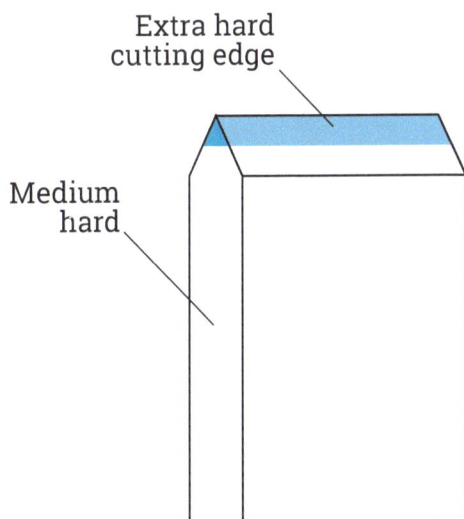

Extra hard
cutting edge

Medium
hard

Figure 2.7 - Hardening of the rule's cutting edge

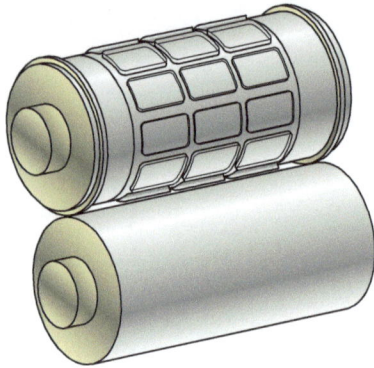

Figure 2.8 - Illustration shows the principle of rotary die-cutting

Materials other than pressure-sensitive are also now being printed and die-cut on label presses by an increasing number of label converters. In particular, forecasts of strong demand for more convenient packaging solutions for processed food, fast-moving consumer goods (FMCG), sandwich packs and pharmaceutical and medical products are driving moves to new rotary converting solutions, including innovation in die cutting technology. Both solid rotary die stations and in-line processing with flexible dies are therefore in common usage and combine to deliver dramatic improvements in converting speed, material throughput and process efficiency in an ever-wider range of applications.

Solid rotary tools are manufactured in various grades of hardened steel (described earlier) – determined by the application and the process used to engrave the die. They can be ground/milled on computer numerically-controlled (CNC) machines or by using electronic discharge machines (EDM). Both have an accuracy in range of 0.0025 mm / .0001".

CNC milled rotary dies. The process of producing solid rotary dies using CNC machines to create the cutting lines starts with the sawing of a tooling steel bar and blank for the gear, followed by turning the shaft, bearer and body of the cylinder. Next up then is the milling of the cutting lines and cutting the gear. The cylinder may be surface hardened prior to machine or hand finishing or after

to make it more durable. With the newest technology, many applications allow for tools to be machine sharpened. Doing so allows for even more consistency in finished surfaces.

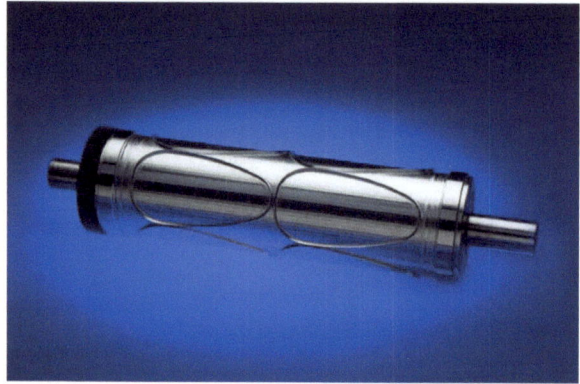

Figure 2.9 - Solid CMC milled rotary cutting die from RotoMetrics

Surface treatments can be added to enhance resistance to abrasive materials, inks and coatings. Non-stick coatings offer a very effective means of controlling adhesive transfer and allowing for the cut of exposed adhesive products.

This finished cylinder will be measured and tested against specifications for quality and performance.

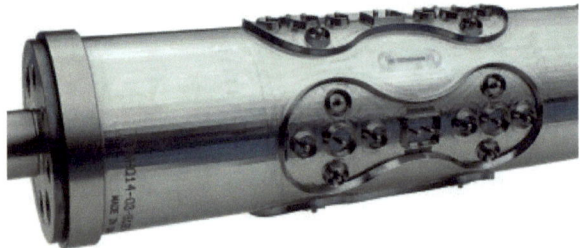

Figure 2.10 - EDM rotary cutting die. Source: RotoMetrics

EDM dies. These are rotary cutting dies produced using electronic discharge machining (EDM) by eroding the non-required steel and leaving the cutting lines. The hardening process takes place before the EDM process, thereby reducing possible distortion of the cutting lines/image.

As with solid rotary dies, the manufacturing process starts with sawing the tooling steel and blank for the gear, turning the cylinder and gear and cutting the gear. Vacuum hardening then takes place before milling the carbon electrode and spark eroding the cutting lines in the EDM machine. Cylinder finishing and testing for quality and performance again take place.

EDM dies provided additional durability for metal-to-metal applications and were also effective for long-run pressure-sensitive jobs, as well as for cutting intricate shapes and designs. Newer technology, though, has the gap closing on these differences between CNC and EDM and the chrome plating of a CNC die can outperform an EDM pressure-sensitive applications

For the growing numbers of converters using special types of CNC or EDM rotary tooling dies that offer waste removal without contamination of the web there are a number of rotary tooling materials management solutions now available from various tolling suppliers, including:

- Pin Eject dies remove or eject the cut part from the die cavity with pins.
- Air-eject and focused Air dies utilize air to eject the waste from the cavity during the converting process.
- Vacuum dies use vacuum to remove the waste through the cavity and out from a center drill in the die.

These various special rotary die-cutting solutions are described in more detail in Chapter 7.

Flexible dies are thin, flexible steel die-cutting 'plates' for use on magnetic cylinder bases. They are produced from specially formulated steel ranging from 0.5 – 1.5 mm in thickness and are then mounted onto a magnetic cylinder. Flexible dies are lower in cost than solid dies and the economics of use become more attractive as the complexity of the label shape

increases. Their life is much the same as for a solid die, providing cleanliness, setting, anvil condition, adhesive and label design are properly controlled.

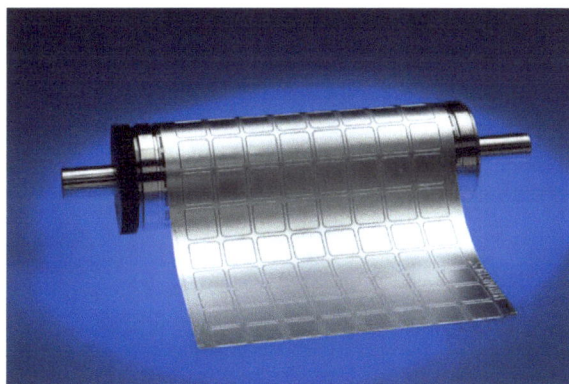

Figure 2.11 - RotoMetrics flexible die

Production of flexible dies involves the plotting of an image directly on the die material so that background material can be removed through chemical etching. CNC mills create the required cutting or creasing lines. Optional processes or features for flexible dies include back grinding, chemical de-burring, laser hardening and coating (non-stick, durability, etc). Before despatch, flexible dies will be tested by the manufacturer on the specified substrate material in a cutting unit.

Flexible dies can be ordered to provide solutions for all kinds or converting requirements, including dies for short, medium and long runs, a wide range of filmic face materials, thin film liners, abrasive substrates and thermal transfer materials, as well as for folding cartons.

Magnetic cylinders used with flexible cutting dies are made from alloy or stainless steel and have a series of permanent magnets glued around their periphery which are used to hold the flexible dies in place. The magnetic cylinders (see Figure 2.12) will fit on to any press that takes rotary dies, and there are no size limitations outside of those relating to the press dimensions. They are installed in exactly the same way that conventional rotary dies are installed.

This can be done by the regular press operators or maintenance staff.

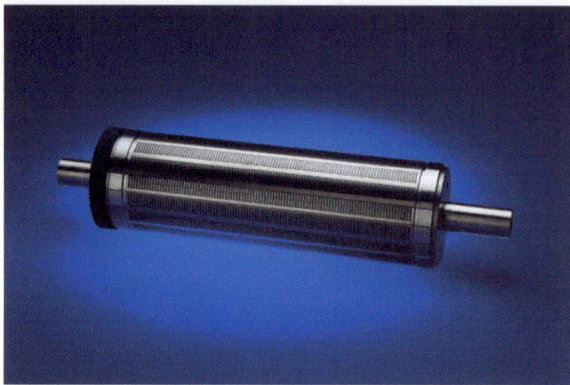

Figure 2.12 - Illustration shows a RotoMetrics magnetic cylinder for use with flexible dies

When using magnetic cylinders, press operators will need to become aware of the high standards of cleanliness required, together with an appreciation of the careful handling required to avoid bending or damaging the plate. The skills required for mounting a flexible die are similar to those required for mounting printing plates. More information on magnetic cylinders is provided in Chapter 7.

DIE-CUTTING UNITS

Whatever the type of rotary cutting – using solid or flexible wrap-around dies – the die-cutting process is basically the same. A web of material will be fed through the press into a die cutting station which holds a rotary tool that will cut out shapes, make perforations or creases and can even cut the web into smaller parts.

A series of gears will force the die to rotate at the same speed as the rest of the press, ensuring that any cuts the die makes line up with the printing on the material. The printing machines can even incorporate multiple die cutting stations, making it possible to die-cut the web in different steps.

The cutting cylinder, either a solid rotary die or a flexible die mounted on a magnetic cylinder, is held in

the die-cutting station or unit in a stable position, at the required cutting pressure, so as to perform the cutting operation. Cutting through the face material and adhesive takes place using the cutting die, cutting against an anvil and backing roller supporting the release liner and cutting process.

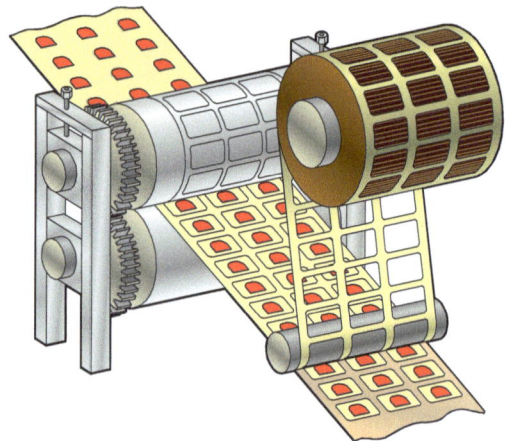

Figure 2.13 - Rotary die-cutting unit showing cutting cylinder, anvil roller and matrix waste removal

The cutting and anvil cylinders are typically stacked on top of each other in a conventional cutting unit. See Figure 2.13. There are some die-cutting stations, though, that load the die horizontally against the anvil. The repeat size of the cutting cylinder will vary from job to job. A cutting unit is generally constructed so that the cutting cylinders are relatively easy to replace.

The resulting matrix waste after die-cutting is removed using either an in-line matrix removal unit on the press (again see Figure 2.13 or extracted through a vacuum waste unit.

Since there is no stop and go-motion in a full rotary cutting process, the production speed can be greatly improved compared to previously mentioned flatbed die-cutting operation. In its simplest configuration there are no adjustment options for the cutting depth like in a flatbed die-cutting operation.

The facility to make very fine adjustment is an important requirement of modern rotary die-cutting units and they are therefore fitted with pressure measurement gauges (see Figure 2.14 and Chapter 8) which enable the operator to monitor the pressure being applied to the die and avoid any overheating of the bearers and subsequent damage to the die. It is also important that all bearers are adequately lubricated throughout the print run.

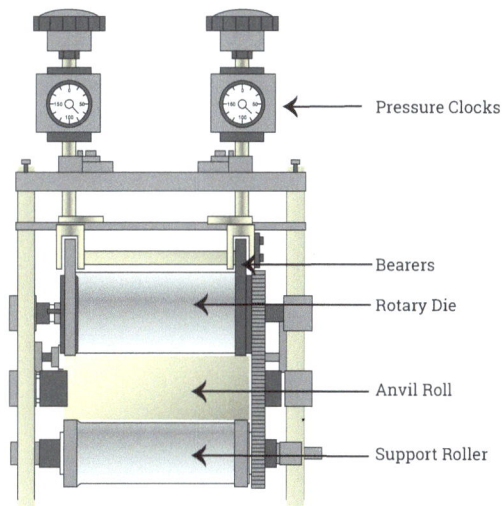

Figure 2.14 - Diagram shows the various components of a modern rotary die-cutting unit

Some label shapes can prove difficult during the matrix removal process and careful consideration needs to be given to the job layout to ensure that the removal of the matrix is trouble free and enables a consistent running speed. A number of presses are equipped with a clutched, driven roller ('capstan') placed just above the matrix take-off point. The pull speed of the roller can then be adjusted to increase or decrease the tension applied to the matrix and allowing the operator to have more control of the matrix removal process.

Modern magnetic flexible dies can be used to cut almost any conceivable narrow web printing product perfectly – from simple rectangles to multi-layer booklets.

As a guide, the advantages and disadvantages of rotary cutting technology can be summarized as follows:

Advantages
- Running speeds are greater than that achieved with flatbed die-cutting
- There is no need to adjust the cutting depth
- It is easy to monitor the pressure being applied and avoid bearer overheating or die damage
- Cutting cylinders are relatively easy to replace

Disadvantages
- The repeat size of the cutting cylinder will vary from job to job
- Solid rotary dies will be more expensive than flat bed dies

SEMI ROTARY/INTERMITTENT DIE-CUTTING
In the semi rotary (intermittent) die-cutting process, the web travels through the machine with a forward and backward movement and the cutting cycle only takes place as the web travels forward. The cutting unit has a 'fixed' magnetic cylinder determined by the press manufacturer's design. (19" or 25.5" are very common).

Figure 2.15 - A semi-rotary die-cutting station with full rotary option. Source: ABG International

The actual cutting length is defined by the amount of web travelling through the press and this 'pull' distance is controlled by the translator feed system, which operates via servo driven cylinders and nip rollers positioned at the infeed and outfeed sections of the machine.

Although the distortion of the flexible dies is based on the circumference of the 'fixed' magnetic cylinder, the total cutting image on the flexible die can be smaller. For the tooling supplier, it is important to know exactly the machine type. This information is necessary for choosing the right template (indicating registration marks/positioning holes/etc.) for the flexible die to be produced.

The speed of this die-cutting process is significantly slower than full rotary die-cutting and is mainly used for short runs and the finishing of digital labels.

MALE/FEMALE FLEXIBLE DIES
These are a relatively new development using Rotary Pressure (RP) technology. This technology is quite different from traditional crush cutting in which a sharp cutting edge cuts through the material against a solid anvil. In RP cutting (See Figure 2.16), the dies have flat cutting edges and burst the material as they squeeze from both the top and bottom.

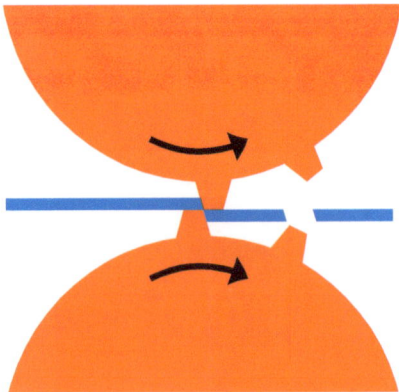

Figure 2.16 - Rotary pressure cutting explained

Instead of using segmented rotary dies, male-female flexible dies (see Figure 2.17) have been introduced for producing folding cartons, drinking cup walls, etc.

The development of RP male/female flexible die technology responds to customer demand for more flexibility in production, reduced set up times and considerable cost savings in tooling within the

packaging industry.

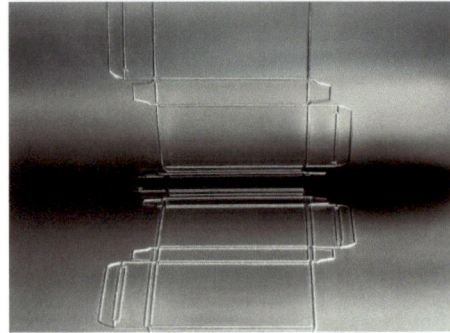

Figure 2.17 - Male/female flexible dies. Source: RotoMetrics

LASER DIE-CUTTING – LIGHT AMPLIFICATION BY STIMULATED EMISSION OF RADIATION
Being able to change the shape of labels on demand and not having to purchase expensive tooling for each shape or size of label – and often delaying the converting line – has meant that laser die-cutting has long excited the label converter. With today's technology now far superior to that of just 15 or so years ago, it has become faster and more efficient, has a higher cut quality and, if used for the right kinds of applications and markets can certainly both reduce a converter's costs and create value-added opportunities for many very short run applications.

Laser die-cutters today are able to take any vector-based digital image – perhaps generated on an Esko system – and import this into the cutter's operating software so as to generate the job set-up within a few minutes. Hence the term digital die-cutting. Quite simply, the laser cutter uses programs developed from the step-and-repeat function of label origination to guide the cutting head around each separate label profile. Digital label printers with digital laser cutting can therefore move from the artwork stage to a finished printed and die-cut label within a very short period of time.

In operation, the laser cuts through the face material – either paper or filmic – and stops after the adhesive layer. Laser cutting like this is not cutting in a traditional sense. The laser is either melting, burning or evaporating a thin line through the material.

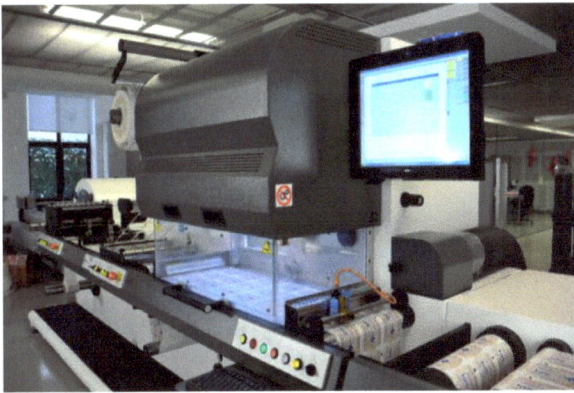

Figure 2.18 - Illustration shows a SEI spa Label Master modular digital finishing unit with laser die-cutting

The laser beam is created by the introduction of gas and electric current to a sealed chamber. As the electricity breaks down the gas, energy is released and resonates between mirrors within the chamber. As it resonates it increases in intensity and at its optimum is released through a partially transmissive mirror. The beam is then directed to a focusing lens and is further intensified. At this point the laser beam becomes a usable cutting device.

The laser light is especially hazardous to the human eye and needs a protective shield to prevent any direct or reflected beam that can hit a person. Because the laser is removing material in the form of fumes, it will require extraction. Laser cutting can cut almost all traditional materials depending on how well the material absorbs the laser energy.

Job changeover from one image to another has also been addressed and is faster with the latest laser cutting technology since the next job can be downloaded while the current job is still running. Larger die cut patterns can be loaded plus static pattern cutting for testing small lots of material samples. The images can be easily stepped across the web, the material cut type and cut path selected, with the cutter then ready to go.

Some laser cutting systems today have manual and prompt pattern changes with the ability to store millions of patterns and can also be networked to an art department or the Internet with newly created cutting patterns being immediately available to the operator.

Depending on the supplier, software may be incorporated in the equipment so that the operator can import or create the die line pattern, edit the die line pattern, and then test the die line on a virtual machine before going to actual production of the finished product.

The most important consideration in laser die-cutting is not the linear cutting speed. It is the actual speed that the web moves through the machine, which is governed by the complexity of the artwork and the ability of the software to optimize cutting. Undoubtedly, the best laser cutters are able to automatically optimize the cutting sequence through the software to produce the maximum web speed. Advantages and disadvantages of laser cutting can be summarized as follows:

Advantages
- The technology offers 100 percent savings in cutter tooling. No dies are required, whether flat, rotary or flexible.
- Tool-less non-contact laser production offers multiple depth cutting possibilities and can include kiss-cutting, thru-cutting and perforation in one pass.
- Converters using laser cutters estimate that the savings in set-up waste can be as high as 60 percent.
- Pattern changes are simple, requiring only edits to drawing files without any downtime for set-up or die creation.
- Lasers today can be used to cut all shapes and sizes.
- With few exceptions, lasers can cut most types of substrates.
- Laser cutting can also be combined with other laser processes such as perforating, scoring, kiss-cutting, etching and ablating.
- Laser cutting can be used to convert difficult materials, such as abrasives and adhesives, with ease.

Disadvantages
- Cutting speeds will depend on a number of variables, including material thickness, the amount of cutting required, the amount of small radius curves, the complexity of the cutting shapes, the number of labels across the web, and the amount of jumping between features.
- The more complex the shape to be cut, the slower the cutting speed.
- Higher energy costs than conventional die-cutting.

Put together, laser cutting can play a valuable part in a modern label converting plant, working with conventional and digital printing outputs to offer materials and cost efficiencies, reduce waste, provide added-value opportunities, sequential coding and numbering, and more complex shapes and lengths. The technology has a valuable role to play in the future and more label converters will be evaluating the benefits and opportunities when drawing up their investment plans.

Chapter 3

Optimizing the die-cutting process

The best die-cutting results are not just reliant on the quality of the cutting die. There are many other factors involved, not the least of which is the nature of the material being converted. Other relevant or important factors can be the way the press or cutting unit is set up and run by the operator, including things like the web tension of the substrate, the running speed of the press, the die-cutting pressure, the stability of the cutting unit, or the removal of matrix waste.

The possibility of cylinder deflection or misalignment can also be an important factor, as can anvil roller diameters, cylinders losing traction, sturdiness of the side frames, and friction in the cutting unit creating too much heat. So let's look again at the basics of a rotary die-cutting unit (Figure 3.1) which, as can be seen, consists of:

- side frame
- bearing block
- pressure bridge
- magnetic cylinder or rotary die
- anvil roller
- support roller

These elements are indicated in the diagram, Figure 3.1. The correct setting and adjustment for all these elements of the cutting unit, both on their own and in relation to each other, are ideally required for the optimum results. Things can still go wrong, though, and not provide maximum cutting performance.

In order to make an accurate and successful cutting separation of the converted material during its process through the die cutting unit, there is a range of parameters that need to be considered as

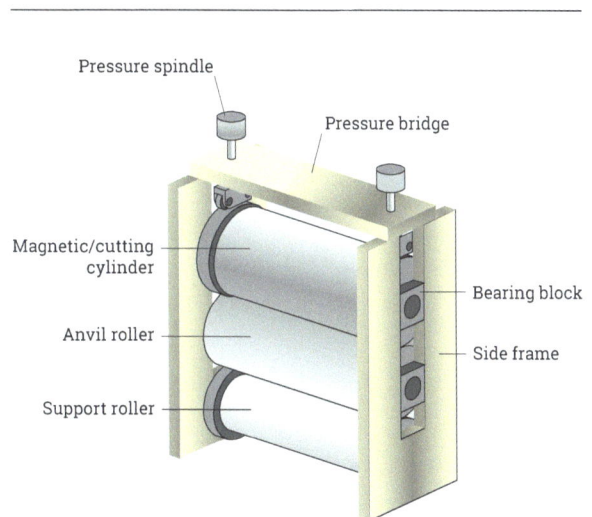

Figure 3.1 - A rotary die-cutting unit

influencing the process. Some of the most important of these parameters are examined in this chapter, namely:

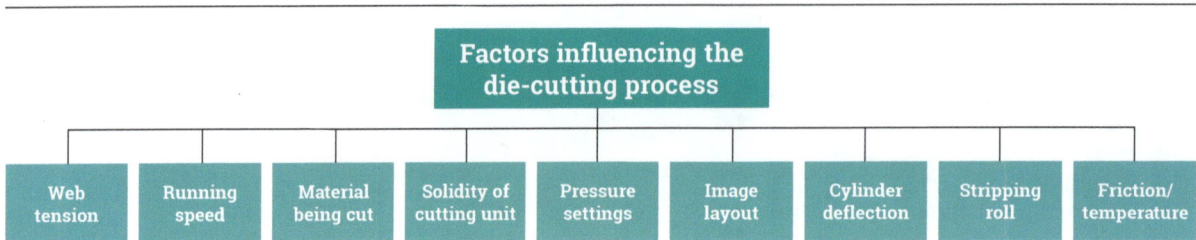

Figure 3.2 - Factors influencing the die-cutting process

Fortunately, there are test methods and procedures that can be used to assess the performance of many of these parameters. The necessary steps can then be taken to eliminate or minimize their effects.

One simple test procedure is the use of an ink stain on the liner material using a broad tipped permanent marker. A full cylinder repeat of the liner material is typically ink tested.

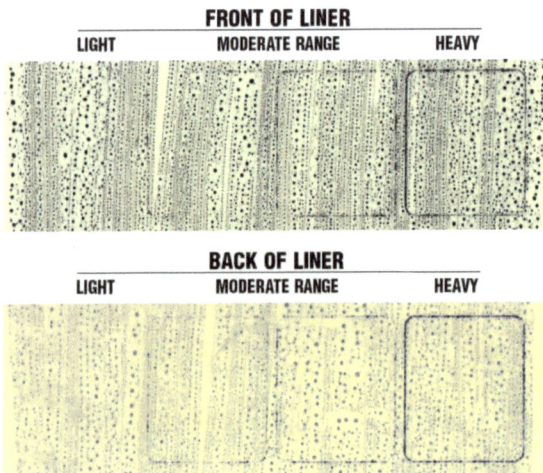

Figure 3.3 - Shows examples of liner impression 'liner strike'

For automatic label dispensing it should be possible to see an outline of the shape in the front or back of the release liner – depicted as a light liner impression or liner strike (Figure 3.3). However, no

fluid or ink should be absorbed by the paper fibers themselves. This can be observed by viewing the liner from the back side.

Labels that are semi-automatic or hand applied can have some slightly visible ink penetration when visually inspecting the liner.

Official test procedures have also been drawn up by some of the main label industry associations. In particular, the following Finat Test Methods are available:

- **Finat Test Method FTM 23a Die strike for paper**
  This test allows the converter to assess the degree and consistency of the die strike and die cutting during the conversion process. The method can be used during press makeready to assess the condition and settings of cutters, to prevent label dispensing failures or web breaks during high speed dispensing. The test is applicable to paper based liners
- **Finat Test Method FTM 23b. Die strike on clear filmic liners**
  This test is used for the evaluation of backing damage or marking to the liner that may be caused due to kiss cutting via a die.

Using the basis of these test methods and recommendations provided by cutter and tooling manufactures enables guidelines to be established for the key parameters. These are outlined below.

WEB TENSION

Web tension issues are most commonly found when

filmic materials are being converted. The filmic materials, for either mono-web or laminate structures, will exhibit elastic behaviours under tension. Increasing tension will eventually reach a point where these films will deform irreversibly and can even cause them to break. Web tensions in pressure-sensitive filmic labeling however, will not generally get to this point. If the liner can be tensioned successfully without breaking, the issue of die strike should not be a problem. The evaluation of any backing damage or marking to the liner due to die strike can be assessed using Finat Test Method FTM23b: Kiss cutting of filmic liner.

It should be noted that web tension alone in reel fed printing and converting does not mean anything on its own unless this is in relation to the elasticity of the material being processed. For a specific 'pull' on a web, both the width and the calliper (thickness) of the web material has an influence on how elastic and 'stretched' the material may become.

For example, if the web width is doubled for a given brake force, the relative web tension will be half. The same goes in theory for the thickness of the material if this is uniform. However, this is not necessarily the case for pressure sensitive material since it consists of at least three separate layers: release liner, adhesive and face material.

In effect, the way web tension has an influence on a specific web material is that when it is being stretched due to the brake force, both the width and the thickness of the material will theoretically decrease, since the volume is constant. This behaviour is rather like how a rubber band acts under tension, only the web material is much less elastic than the rubber band.

Since die-cutting is usually performed on a web under tension, the cut-out label will shrink slightly in length and grow in width when the material is in a relaxed state again. By how much depends on the elasticity of the material being converted and the tension of the material during the die-cutting process.

SPEED
Web speed can have an impact on the die-cutting performance, particularly if the die cutting force is not uniform in the longitudinal direction.

For example, in a web with a number of square labels and a small gap around each label, the cross-cutting section of the die requires a sudden drastic increase in cutting pressure or force (see also under Pressure/Force and under Image Configuration). This phenomenon is commonly manifesting itself as bounce and is caused by the impact energy, where speed is playing a major role.

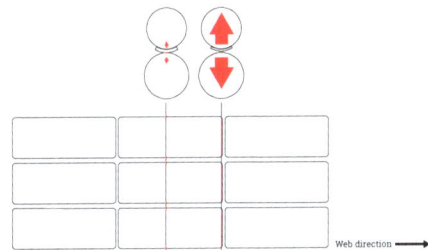

Figure 3.4 - A sudden increase in pressure across the cross-cutting section of the die manifests itself as bounce

MATERIAL/SUBSTRATE BEING DIE-CUT
As previously discussed, pressure-sensitive materials are a laminate construction consisting of at least three layers: the face material, adhesive and the backing or release liner. If a die should cut into the liner excessively this could reduce the tensile strength of the web sufficiently to cause a web break. As such, damage in the transverse direction (TD) across the web can be more of an issue than in the machine

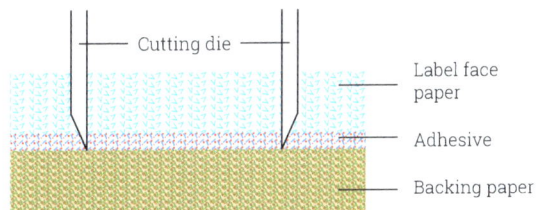

Figure 3.5 - The die-cutter needs to cut through the label face material and adhesive, but not the silicone coating of the backing liner

direction (MD) along the web. To produce the correct

tooling, the die manufacturer needs to know what type of material is going to be converted, in particular, the thickness or caliper of the release liner and the type of liner being used, e.g. glassine, kraft, PET, PP.

During the 'kiss cutting' process, the cutter edge needs to cut through the face material and adhesive onto the liner, but without damaging the actual silicone layer of the liner (Figure 3.5). This means that the tooling supplier will have to determine how deep the cut must be in order to cut through the specific material being converted, but not cut into and damage the silicone coated liner (die-strike).

Die-strike through filmic liners can be influenced by several parameters. It is recommended that the following parameters are carefully checked if die-strike problems occur:

- the solidity and robustness of the die-cutting unit
- the tool diameter should be adjusted according to the width of the printing equipment
- the tolerance between magnetic and anvil cylinder
- the settings of the flexible die (wear-ness - height profile of the die)
- the consistency of the liner thickness
- the temperature at which the die-cutting operation take place (influence of UV light on the film and adhesive softness).

To eliminate or minimize these problems it may also be feasible to adjust the cutting angle to improve the die-cutting operation, verify the tension of the web (avoid too high a tension), and strip the matrix immediately after the die-cutting operation to avoid recovery of the adhesive between the die-cutting and the stripping steps. Consider cooling down the laminate before the die-cutting operation.

Depending on the cutting angle, the sharpness of the cutting edge, and the specific material properties of the face material, a certain force will be acting on the release liner at the precise moment of cutting. The release liner is compressible like any other material and will compress from the force applied by the cutting tool. How much the release liner will compress depends on the thickness and the properties of the liner.

FRICTION/TEMPERATURE

It is not always realized that increasing or fluctuating temperature can have an effect on die cutting tools and more specifically on the temperature of the contact points between the anvil cylinder and die-cutting cylinder. Temperature should therefore be regularly monitored.

The force applied to the cylinders in order to keep them together during a die-cutting process leads to both friction and compression. This takes place in the contact points between the anvil and the cutting cylinder when the cylinders are rotating.

In most cases, the force from the lead screws is transmitted via the bearers of the cutting cylinders. Because of the compression that is caused by the force holding the cylinders together and the friction that is caused by the rotation of the cylinders under load, heat will build up in these contact points.

How much heat is generated depends on a variety of factors such as the load on the cylinders, the width of the bearers, the diameter of the cylinders, how fast they are rotating and how well they are lubricated. If the build-up of heat cannot escape as quickly as it is generated, then the temperature will increase on the bearers of the cutting cylinder and locally on the anvil cylinder.

Any heating-up of the bearers of the cutting cylinder will ultimately cause the bearers to grow in size due to heat expansion. As long as the die-cutting frame support is capable of giving sufficient resistance against the increase in force, the increase in diameter of the bearers will result in an increase in force between the cutting cylinder and anvil roller. Over-pressuring the cutting die can increase the wear to the blades and shorten the die life. This heat and friction can cause premature wear on the bearers of the cutting die as well. It may eventually cause problems with maintaining the correct gap between the two cylinders.

For this reason, it is important not to apply more force than necessary through the lead screws and also to keep the contact area between the anvil cylinder and the cutting cylinder clean and lubricated, otherwise problems such as accelerated wear of bearer rings, press rollers and anvil roller, or even losing control over the die-cutting process, may occur. To avoid such issues, the pressure should be reduced down to the necessary level at the right time so as to prevent excessive heat build-up and cutting into the liner.

Figure 3.6 - Diagram shows the main elements of a rotary die-cutting unit

SOLIDITY AND STABILITY OF THE CUTTING UNIT

One of the key parameters in the die-cutting process is the actual die-cutting unit itself (Figure 3.6), which supplies the required frame for suspension and support of the cylinders during the cutting process and to also provide the necessary stability and resistance to the cutting forces. In essence the cutting unit is responsible for supplying sufficient resistance and stability against all the internal forces created as a result of the cutting process.

The **station frame** and **side panels** of a cutting unit are there to keep all components involved in the die-cutting process in their specified positions. They should be able to absorb all occurring forces safely and vibration-free. Both the side panels and the (pressure) **bridge** should be adequately dimensioned, as otherwise this could lead to instability in the cutting unit and contribute to potential cutting problems.

Optimized cutting systems feature a force transfer from the bridge through pressure adjustment **jacks** or

gauges with sturdily dimensioned threaded rods. Pressure is transferred from the intermediary pressure truck via the fitted pairs of roller bearings to the bearers of the cutting die or magnetic cylinder.

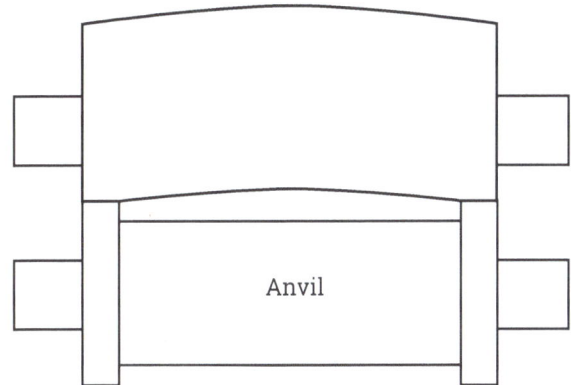

Figure 3.7 - Pressure to the shaft may lead to bending or flexing of the magnetic cylinder

The **pre-load pressure** attained when setting up and the running pressure must be greater than the material resistance when cutting across the web. Therefore, the level of pre-load will depend on the material being cut, the working width of the machine, the label shape and the design features of the cutting lines.

The cutting cylinder (a solid or flexible tool on a magnetic base) is pressed onto the anvil cylinder by pneumatically or mechanically operated pressure systems. Pneumatic force should be used only to control an adjuster cam or toggle lever. It should never be exerted on the press unit, magnetic cylinder and anvil roller, as otherwise vibration may occur.

Mechanically operated systems are divided into systems which are able to apply pressure to the shaft of a magnetic cylinder or die, which may lead in extreme cases to bending or flexing of the magnetic cylinder (Figure 3.7) and in extreme cases to cylinder shaft breakage. Flexing may also occur when using small diameter dies.

Commonly, presses today apply pressure to

bearers on the die. Force transmission is via the pressure bridge.

There are also systems which operate without bearer rings, which means no pre-load is permitted between the cutting die and the anvil roller. The magnetic cylinder or anvil roller is mounted in conically adjustable bearings or is pressed against the resistance by springs. These systems strongly depend on the perfect condition of the bearings that must not have any play, and an adequate dimensioning of the shafts. Such a system is not recommended for the converting of pressure sensitive labels.

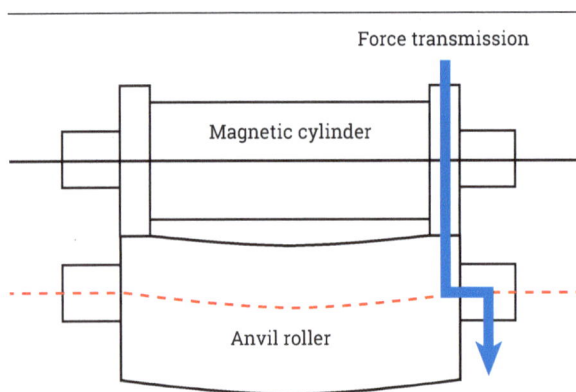

Figure 3.8 - Deflection of anvil roller

Anvil rollers also play an important role in the die-cutting process. Anvil rollers that are too small in diameter and are mounted in the machine side panels without benefit of a support roller underneath pose a risk of deflecting due to the necessary pressing force exerted by the cutting cylinder. See Figure 3.8. The consequence is less pronounced cutting towards the middle of the web.

As was earlier seen in Figure 3.6, modern die-cutting units are likely to contain a **built under** or **support roller** which is positioned under the anvil roller. This roller allows the perpendicular transmission of force from the pressure bridge to the support roller with the aim of preventing bending of the anvil roller due to pre-loading of the magnetic cylinder.

Inadequately dimensioned (too small diameter)

anvil rollers may possibly still deflect despite the extra support from the support rollers especially at the point of cutting a horizontal line.

PRESSURE/FORCE

This refers to the force applied through the two lead screws mounted in the die-cutting unit for both semi- and full rotary die-cutting units.

In flatbed die cutting, the force is not usually something that can be altered. Only the stroke length is adjustable in order to accommodate for the desired cutting depth.

The force required to perform the die-cutting process depends on many factors, some of which have already been discussed, such as the substrate, the web running speed, anvil size and the various elements of the die-cutting unit. Other factors are:

- the geometry and sharpness of the cutting edges on the cutting tool
- the weight of the cylinders and other component parts if stacked on top of each other in gravitational direction
- the bearer width
- the cutting length in contact with the web
- the cylinder diameter

All of the factors mentioned have an influence on the necessary force that will be required to perform the actual die-cutting. Wear of the cutting tool will over time dull the cutting edges, in turn increasing the required cutting force as well.

The pressure required for die-cutting depends upon the amount of blade that penetrates the material in the same instant. The pressure serves to prevent the die from bouncing, by keeping the bearers in constant contact with the anvil roll surface. When the pressure is inadequate the die will bounce. When the pressure is excessive, it will accelerate the wear of the die, the anvil roll and other components of the station.

When the pressure, required to cut cleanly, exceeds the recommended pressure by 400 lbs. It is recommended the die be retooled. Don't wait until the die stops cutting before sending it to be re-sharpened. This practice results in costly emergencies and can reduce the number of re-sharps that can be achieved.

The force applied through the two lead screws needs to exceed the maximum cutting force required by the cutting pattern, where maximum die line is in contact with the web on a rotation of the die-cutting tool. Two close lines across the web direction will require a much higher cutting force than a single line in the web direction because of both the sudden interference with the web and the actual total length of the cutting line in contact with the web. This will be discussed further when looking at die-cutting image patterns.

Certainly there can be big fluctuations in the required cutting force during a revolution of a cutting tool, again depending on factors such as the cutting pattern. However, it should be remembered that any internal bending of cylinders cannot be prevented by simply increasing the force holding the die-cutting and anvil cylinders together. The force applied through the lead screws must only be applied in order to counter for the maximum force required for the cutting edge(s) to penetrate the substrate.

IMAGE CONFIGURATION

In the die-cutting process there are die line/cutter layouts that are deemed to be more preferable than others. Certainly, the most demanding cut in any rotary die-cutting operation is always going to be that of cross cutting, as shown in Figure 3.9.

In most cases, labels are arranged symmetrically to save space and keep the consumption of materials to a minimum. However, this configuration is less beneficial for the rotary die-cutting process because vertical lines tend to cut more forcefully than horizontal lines and high contact pressure damage the liner material and result in wear and lasting damage to all components in the cutting unit.

Poor image configuration may lead to insufficient separation of the web material in the areas where the higher cutting force is required. If the combined rigidity of the framework of the die-cutting unit and the rigidity of the anvil and die cutting cylinder is not sufficient to deliver the resistance against this sudden increase in cutting force, the elements will deflect away from the resulted force and thus momentarily increase the distance between the anvil cylinder and

the die-cutting cylinder, causing a potential flaw in separation of the processed web material. As discussed, Figure 3.9 shows a wide line getting in contact with the material very suddenly.

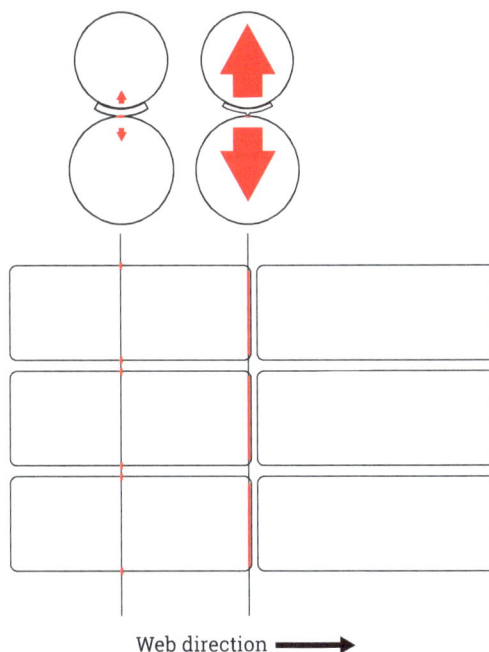

Figure 3.9 - Cross cutting across a label web

A way to counteract this negative effect is by vertically shifting or staggering both the impressions and cutting in the machine's running direction, which reduces the number of horizontal lines and evenly distributes cutting pressure. See Figure 3.10. This possibility is available with, say, Esko's 'staggered cut' software.

Staggered cutting will reduce the pressure required, help to improve waste stripping and prevent die and anvil deflection. If in-mold material, folding cartons, or tag production is being carried out, then staggered cutting will facilitate stacking and more uniform cutting.

While it is not always possible to use this type of solution, the option should certainly be considered whenever it is feasible. If the die line layout is made up of rows and columns of identical patterns, staggered layouts, such as that illustrated, will require less force to cut. The variation of the die line in contact with the substrate material being cut is more homogeneous over one full revolution of the cutting cylinder.

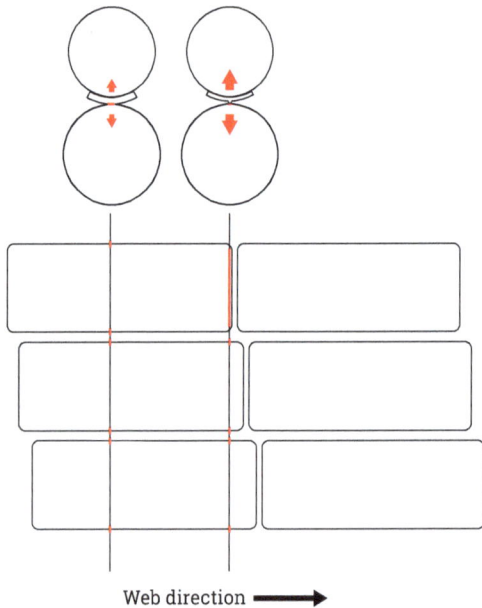

Web direction ➡

Figure 3.10 - More even distribution of cutting pressure using a staggered cutting layout

This type of solution is particularly advantageous when cutting shapes with many horizontal lines. It optimizes die-cutting results and reduces liner damage. The dies last longer as well as various other machine parts. The waste stripping is stabilized preventing web breaks and reducing down time.

It is usually not a problem applying sufficient force through the two lead screws, but the internal deflection of the anvil roller and the cutting cylinder may exceed the desired maximum variation in gap due to the peak in force generated by a wide cross cut.

The force necessary to cut a line across the web is proportional to the width of the line being cut. The deflection of both the anvil and the cutting cylinder will depend on both diameter and length of each cylinder. Since both diameter and length are affecting the maximum deflection to the power of four, it is important that the diameter/length ratio is sufficient to withstand the force from the die cutting process.

CUTTING ANGLE/DIE HEIGHT/CLEARANCE

The geometry of a cutter will vary depending on the material being converted. Based on the construction and thickness of the material being cut – adhesive, liner thickness and type of liner – the three most important die parameters, cutting angle, die height and clearance (drop), can be determined for flexible dies. For rotary dies, both the cutting angle and the 'drop' (distance between tip of cutting line and bearer) are the most important parameters.

The Die Clearance or drop is the difference between the height of the cutting blades and the height of the bearers. It sets the depth of the cut. This is shown in Figure 3.11. To determine the proper clearance, the exact thickness or caliper of the liner material is required.

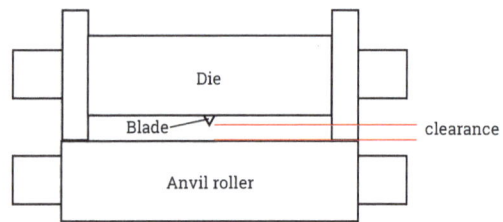

Figure 3.11 - Clearance between cutting blade and anvil bearers

In terms of the cutting angle, this will normally vary between 50° - 75°. The height of the cutting line is measured from inside the pocket to the tip of the cutting edge and will normally vary between 0.38mm – 0.80mm (0.015" – 0.030") for flexible dies and will be 1mm (0.039") or more for solid rotary dies and rule dies. In general, thicker materials require a sharper cutting angle just for the fact that the cutting edge has to travel further through the material as well as for plastic materials such as PE, PET, PP etc.

A narrow cutting angle will, in general, require less

force to penetrate the substrate simply because less area is being pushed through the material. As mentioned earlier, less force also has a positive impact on the deflection of the cylinders, which in turn means the potential for a more even die strike.

The harder types of material, like most plastic filmic materials, generally tend to require a higher cutting force than softer materials like paper and therefore it makes sense to try and compensate for this by making the cutting edge sharper (narrowing the cutting angle). See Figure 3.12. On the downside, a narrow cutting angle also tends to wear out more easily simply because the cutting load is being exposed to a smaller surface area.

The profile of the cutting blade should be fine-tuned to compensate for differences in the thicknesses and elasticity of the substrate being cut. Elastic materials such as polyethylene will require a steeper blade profile with a keener edge than, say, a more rigid material such as paper. Steeper, keener blades also have a lower resistance to wear than thicker blades. This is the reason why there is no such thing as a multi-substrate die.

Using film dies to cut paper may diminish the die life for cutting film. Wear to the blades may not leave the blades with the edge required to cut the film they were ordered for. This could create a costly emergency in production that would erase any cost savings achieved by using one die for several materials.

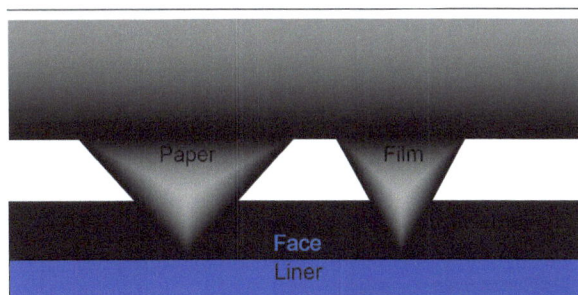

Figure 3.12 - The importance of cutting blade angle for different materials. Source: RotoMetrics

As indicated in Figure 3.12, softer materials like

the different paper substrates will normally benefit from a wider cutting angle because the material is easier to die-cut and the wider cutting angle will generally expand the life of the cutting edge.

The cutting angle is determined mostly by the properties of the face material to be cut. As explained earlier, face materials with high tensile strength like most plastic materials are usually cut with sharper cutting angles in order to decrease the cutting force and thereby also the compression of the release liner during the cutting process. Thicker types of materials often also require sharper cutting angles in order to accommodate for sufficient space inside the die cavity during the cutting process.

For environmental, economic and technological reasons, label stock suppliers have significantly changed the specifications and use of their release liners, such as:

- moving from kraft to glassine liner
- reduced caliper glassine liner
- trending towards thin PET liner such as 36 micron to 19 micron liners (.0015" - .00075") or even thinner

The use of thinner liners will reduce the tolerance in variation for cutting force simply because there is much less compression to work within. Thinner liners require narrower tolerances in tool making in order to get consistent performance and maximum life from the cutting tool.

When looking at the overall thickness of pressure-sensitive materials, the tooling manufacturer needs to determine if the inside pocket clearance (the blade height) is sufficient. For thick materials, the cutting angle might also be affected by this as thicker materials may require sharper cutting in order to preserve the label edge.

MAGNETIC CYLINDER DIFFERENCE

The difference (diff) or undercut of a magnetic cylinder is the difference between the diameters of the bearers and the diameter of the body of the cylinder. This measurement is crucial to calculate the proper flexible die height.

The most commonly used difference is 0.038" /0.965 mm. This would be a clearance on each side (also called an 'air gap') of .019" / 0.48 mm (Figure

3.13). This size however can vary. Cylinder manufacturers can adjust the diff based on the requirements of the converter. A larger 'diff' may be used when a taller plate height is required (such as for thicker materials). If the flexible die manufacturer did not manufacture the magnetic cylinder though, determining this diff is a necessary part of manufacturing the flexible die. The converter can supply this information as provided to them by their magnetic cylinder supplier. It can be measured also. Guessing at the diff though can risk possible errors and reordering of the flexible dies.

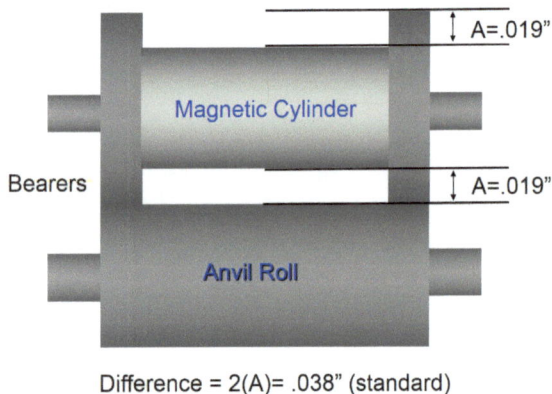

Difference = 2(A)= .038" (standard)

Figure 3.13 - The difference or undercut of a magnetic cylinder

A number of magnetic cylinder suppliers, such as RotoMetrics, keep a database of the differences for all of the magnetic cylinders they manufacture. Each flexible die is custom manufactured to fit a specific cylinder for the most precise cutting results.

The bearers of the magnetic cylinder will wear over time and should be measured every three or four years to determine how much the diameters have changed. In many cases, the magnetic cylinders can be ground to re-establish the original 'diff'.

CORNER SHAPES/RADII
A square cornered label or waste matrix does not release as easily from a liner as a rounded corner will. A corner radius is used to make it easier for

automatic label application as well. It can adhere better to the final product. The term 'corner radius' itself refers to the radius of the circle created by extending the corner arc to form a complete circle. Figure 3.14 shows a corner radius guide.

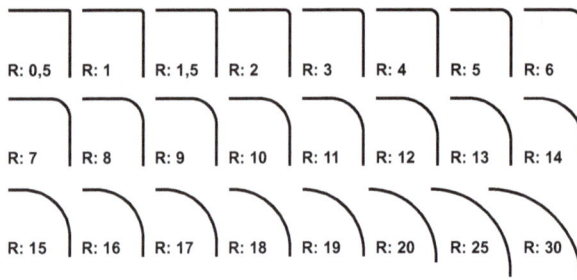

Figure 3.14 - A guide to corner radius. Source: Electro Optic

Square corners may increase the cost of the die and reduce productivity and die life. It is recommended that a corner radius of at least 3 mm is provided. The minimum recommended radius is 1 mm (.0394"). There may be a supplementary charge per corner for radii less than 0.793mm (.0312") due to the added difficulty of manufacturing them.

CUTTING DEPTH FOR LABELS AUTOMATICALLY APPLIED
For automatically applied labels, the cutting blade should burst the face stock and adhesive without penetrating through the silicon coating on the liner. However, die life can be shorter because the die may stop cutting after minimal wear. As mentioned previously, this bursting process is difficult to achieve when cutting very elastic synthetic face stocks or when cutting to a soft thick liner. If the die cuts too deep, though, it can cause the liner to break in the label applicator or the de-lamination of the silicon layer along with the label.

For hand applied labels the blade should burst through the face stock, adhesive and slightly penetrate the liner.

In all cases, always make sure to tell the die supplier how the label will be applied!

CYLINDER DEFLECTION

Mention has already been made in preceding pages to the deflection of die-cutting cylinders or anvil cylinder under pressure and in certain conditions. Such deflection can critically affect die-cutting results. It is not technically possible to apply pressure across a cylinder suspended on two outer points or bearers without the cylinder bending or deflecting to some degree under the pressure load. And as previously discussed, the internal deflection of the cutting cylinders can be critical to the cutting result. Fortunately, it is possible to keep the deflection to a working tolerance if various issues can be addressed as follows:

The combined deflection between both the anvil and the cutting cylinder is what determines an increasing deflection gap. In other words, deflection is not just a phenomenon attributed to the cutting cylinder alone but additionally has a component coming from the anvil roller. During the rotation of the cutting cylinder, the cutting pressure will vary with the amount of cutting edge engaged in cutting. To overcome this, the pressure applied to keep the die-cutting cylinder and the anvil roller together during the cutting operation must exceed the peak in pressure caused by the maximum cutting edge engaged. If not, the two cylinders will make a relative move away from each other.

However, deflection is not the same all across the full width of the cylinders. Since the applied pressure is transferred through the bearers of the two cylinders there will be no deflection (lift) in these contact points. The maximum deflection will occur along the cylinder widths where the distance from the contact points is the greatest for an even distributed cutting force - typically in the middle of the cylinder's body width. See figure 3.15.

Factors which will impact on the amount of deflection in the cutting and/or anvil cylinders are the distance between the bearers, the diameter of the body, the cutting force, and the type of steel used for producing the cylinders.

With the maximum deflection experienced at the center of the cylinder, any deflection will cause a variation in the gap size over the whole width of the

cylinder. If any variation in gap exceeds the tolerance for cutting the material, insufficient or poor separation will be the result.

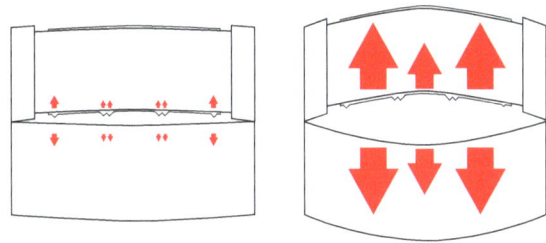

Figure 3.15 - Deflection of cutting cylinder / anvil roller

Since the lowest cutting height inclusive tool wear and deflection will determine when a cutting tool is no longer useful, the D/L (Diameter/Length) ratio can have a dramatic influence on the cutting tool life. For this reason, cutting tools potentially last longer on bigger diameter cylinders than smaller especially if compared to D/L ratios in the critical range.

CYLINDER CONCENTRICITY

Concentricity can be said to occur when two or more objects share the same center or axis. For example, all the features shown below can be said to be concentric.

Usually, designs require that a feature be round as well as concentric like example A below (Figure 3.16).

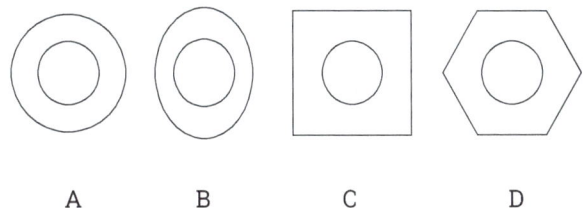

A   B   C   D

Figure 3.16 - Examples of concentric features

A better geometric control is usually circular run-out. Circular run-out controls circularity (roundness) as well as concentricity.

Accuracy in tool making has improved significantly over the years. The range in which a cutting tool will actually cut a material is only a fraction of the thickness of the material. This leaves the tool maker only a very small tolerance to manufacture the cutting tool.

Any impurity trapped between the surface of a magnetic cylinder and the backside of the flexible cutting tool will influence the cutting performance, any impurity between the bearers of the cutting cylinder and the anvil cylinder will have an influence on the cutting performance, deflection of cylinders will decrease the working range for the cutting tool.

Besides supplying sufficient resistance against the cutting forces, the cutting unit is also responsible for the critical aligning of the axes between the cutting tool and the anvil cylinder. If those two axes are not parallel, a heavier cutting impression will be dominant in certain areas of the width of the cutting. This tendency is more critical for cylinders with a smaller D/L ratio (see deflection).

Imagine two cylinders on top of each other perfectly parallel. The contact between these two cylinders will be a line with no possibility to make any rocking motion. The same two cylinders with non-parallel axes, will have a single point of contact with the possibility of rocking. See Figure 3.17.

STRIPPING ROLL

The matrix stripping roll should be available in various diameters and facilitate easy removal of matrix waste. Larger diameter stripping rollers are suitable for stripping materials that can easily tear, such as paper. Smaller diameter stripping rollers are more suitable for converting plastic materials and small labels.

The speed of a machine can usually be increased using blunt stripping knives instead of stripping rolls. Maximum flexibility is needed with regard to the positioning of the stripping rolls in order to facilitate the waste stripping.

The matrix must be guided by way of a closely controllable pull roller to the rewind spindle or to the extraction hopper. The correct web tension is a key aspect of the matrix separation process.

SLIP

This is the relative movement of an object or surface (web) and a solid surface (rollers/cylinders). In web handling this refers to rollers/cylinders losing traction if conditions are not within tolerance. This phenomenon may have an impact on the control of the web-tension during the production process and can even lead to problems with the measurements of a finished product.

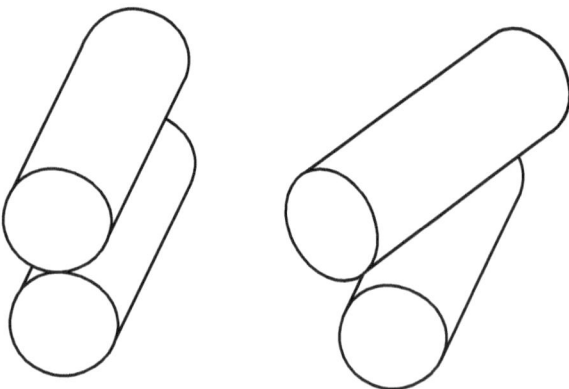

Figure 3.17 - Misalignment of cylinders

Chapter 4

Special tooling for cutting, perforating, hole punching and slitting

While the commonly available flat, rotary and flexible dies can be used for the great majority of cutting applications, there are nevertheless some types of cutting, converting and finishing applications which require specialized types of tooling.

These specialized tools include removable blade sheeter/perforator cylinders, pin and spring eject dies, standard and dedicated air eject dies, shock air eject cylinders, vacuum rotary dies, hole punching rings and score/punching units, even graining and texturing. In fact, quite a range of solutions that may be used by label converters as required. Some of the key solutions can be seen in the flow chart below (Figure 4.1).

Each of the specialized tools itemized in the flow chart will now be examined in more detail

PIN EJECT DIES
Small die cut areas can stay in the die and accumulate to the point of putting pressure on the inside edges of the cutting blades. Internal pressure will cause the cutting blades to break. Pin eject dies were developed as a solution to prevent this build-up.

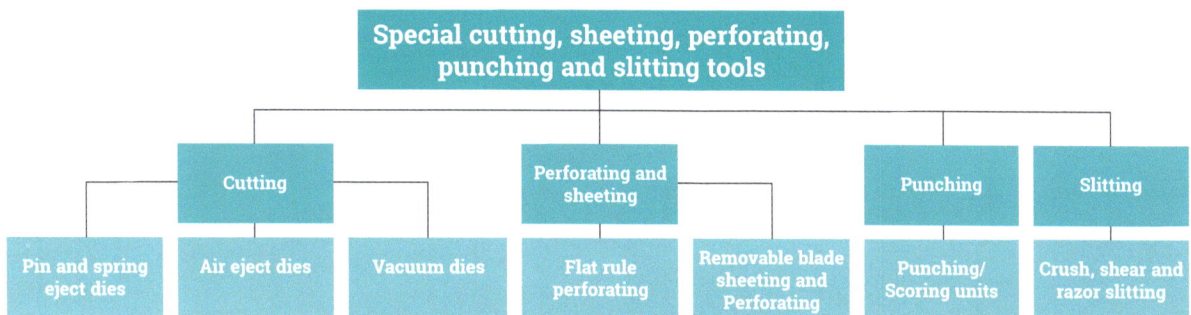

Figure 4.1 - Special cutting, sheeting, perforating, punching and slitting tools

It can also assist in the removal of the waste from the converted web.

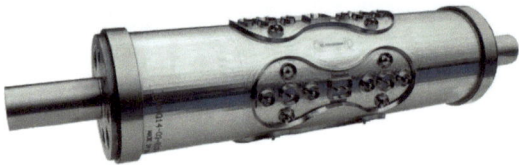

Figure 4.2 - A pin eject die courtesy of RotoMetrics

Pin eject dies contain a special compressible core material that allows the hardened ejector pins to compress into the die upon anvil contact. As the pins rotate off the anvil surface the unique engineered core will force the pins back out of cavity, removing the waste slug from the cutting die. This revolutionary new product (developed by RotoMetrics) has not only improved current manufacturing methods, but has opened the door to new markets that require small die-cut shapes.

The feature provides a proven solution for metal to metal cutting challenges and will also allow the die cutting of small irregular shaped parts that need to be retained on the carrier liner. Adhesive displacement caused by the cutting action of the blades has made this all but impossible in the past. Using this technology, these cavities can now be cut clean and forced back into place utilizing the pin eject system.

Applications include disposable medical products, automotive gaskets, pharmaceutical applications, laser sheets, no waste boarding passes and tea bags to mention just a few.

Figure 4.3 - A RotoMetrics spring eject die

SPRING-EJECT DIES

Spring eject dies are designed to prevent waste build-up inside die cavities and assist with waste removal out of the web. Since it uses spring-loaded pins to eject material out of the web on dies cutting through to the anvil, Spring-Eject Dies do not need compressed air.

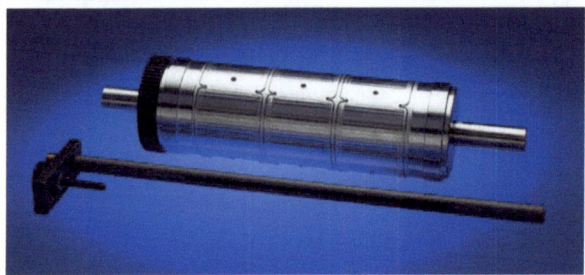

Figure 4.4 - Shows a dedicated RotoMetrics air eject cylinder

AIR-EJECT CYLINDER DIES

Air-eject cylinder dies use compressed air to blow cutter waste out of the cavities during the cutting process, avoiding the build-up of waste inside smaller cutting cavities. The force of the air blowing through each individual hole is determined by the number of holes on the cylinder and the capacity of the air compression being used. Different types of air eject dies are available according to specific requirements, namely:

- **Standard air eject** dies use compressed air forced through a shaft bore to blow air out of all holes drilled in the various cavities. The force of the air is determined by the capacity of the air compressor and the also the number of holes on the die. For these types of cylinders there is little control over the waste ejection.
- **Dedicated air eject** cylinder dies make use of a special probe that fits precisely into the shaft bore and enables compressed air to be directed only to one row of labels/cavities across the web. See Figure 4.4.
- **Multi-port air eject** cylinder. Developed as an alternative to the more expensive male and female cutting system for long runs and the solid rotary die-cutter with rubber inserts and

without any waste control, shock air systems make use of so-called air forks to blow air into channels. These air channels, are linked to bores in the cavities.

Figure 4.5 - RotoMetrics Multi-port air eject cylinder

Air eject cylinders are usually used in conjunction with an extraction or waste collection unit which has been specially designed to collect the small pieces of material blown out of the die in order to avoid the waste collecting in the gears or on the surface of the anvil cylinder.

Die-cutting of the holes and blowing out of the waste into a stainless steel vacuum box or waste collection device is done simultaneously. Finally, the collected waste will be sucked away by an industrial vacuum cleaner.

VACUUM (THROUGH SHAFT BORE) ROTARY CYLINDERS

Vacuum rotary die cylinders provide an alternative to air eject cylinders. Instead of blowing the waste material out, the vacuum cylinders actually suck the cut-out waste particles into the die. It is then removed through the journals with the use of a vacuum extraction system.

Quite simply, vacuum dies feature a series of interchangeable punches which pick up waste from the web and pull it through the die journal with the assistance of a vacuum attachment. They are specially designed to cleanly remove even the smallest pieces of waste from the web, and are said to be the cleanest way of removing waste from the web when die-cutting.

The size, shape and number of cavities to be vacuumed all determine whether or not this type of product should be used. Vacuum dies also require the use of a vacuum waste assist block. Extra punches may be ordered with the tool and operators can quickly and easily replace any punch in-house, minimizing downtime.

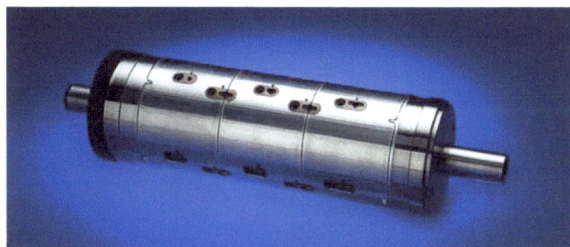

Figure 4.6 - Vacuum cylinder courtesy of RotoMetrics

Vacuum systems are also available for use with specially designed magnetic cylinders with an air system incorporated. In this case, the flexible die has holes in it, distributed across the plate as to line up with the holes in the magnetic cylinder, allowing the suction (or even air expulsion) of the die cut waste.

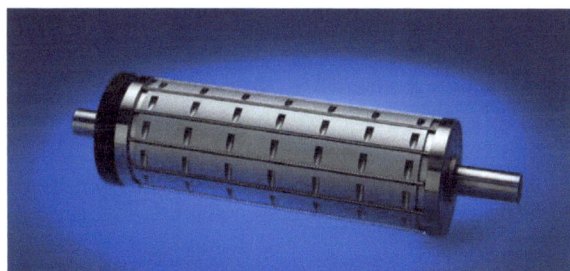

Figure 4.7 - Removable blade perforator or sheeter courtesy of RotoMetrics

REMOVABLE BLADE SHEETERS AND PERFORATORS

Removable blade perforators and sheeters come in a number of different formats, depending on the particular manufacture, and may be in the form of a removable blade or a floating blade – each of which can be made with multiple, precision-milled slots at equal or special locations around the roll. They are designed for cutting across a web to produce items such as A4 sheets, or to provide perforation. These lines of very small dashes are punched into a label surface enabling various parts of the perforated label to be separated from one another by simply tearing off along the dotted perforated lines.

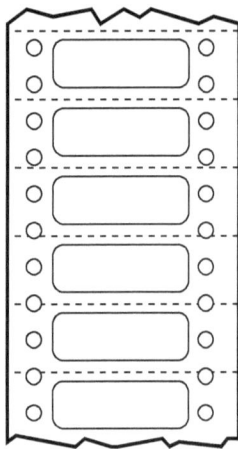

Figure 4.8 - Sprocket hole punched and perforated computer labels

The blade seats are milled and precision-cut to ensure perfect location. The clamping bars and counter clamps ensure that blades are fixed securely. Fast and reliable blade assembly is guaranteed every time. For sheeting, blades are usually made for through cuts, but other heights can be made upon request.

The sheeter or perforator shown in Figure 4.7 is made for through-to-anvil (metal-to-metal) cutting only and does not require the placement of shims under the blades as in some manufacturers' units. It does, however, have shim stock to the side of the blade to accept the pressure of the screws holding the blade in the slot. Any distortion from tightening the set screws is transferred to the shim then, instead of the blade. Slots can accept either a perforating blade or a sheeter/scorer blade. Perforating blades are also available in several standard cut and tie combinations, or can be custom made. Minimum distance between

blades varies depending upon the diameter of these units.

Perforating and hole punching (described later) are often used together and have of course, long been used within the continuous and fan-fold sprocket hole punched business forms and computer label manufacturing sectors to carry labels through printing devices (Figure 4.8) using tractor-, pin- or sprocket-fed systems. Perforations can be produced to be vertical or horizontal, and may go through the label only, through the carrier between the labels, or through both the label and the carrier.

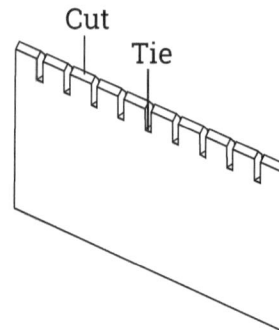

Figure 4.9 - Perforating blade showing a typical cut and tie pattern

Perforations are available in a wide range of strengths, full or partial, depending on the size of the cut to the width of the connection (called the 'tie'). The number of ties per inch and the width of the tie and the cut are based upon the weight of the liner, the type of printer used, the type of burster, and the kind of handling that the label will be exposed to.

Depending on the manufacturer, custom perforations may be available by size or length of perfs, or the number of dashes, reducing or increasing in size, and the quantity of perfs required.

Apart from continuous and fan-folded labels, tickets, tags and business forms, perforated labels are used for applications such as coupons, gift certificates, pharmaceutical labels, business, and industrial applications.

Perforated stickers are also often used for pull out security tags that need to stay intact until the stub tag is pulled off. Such labels can be used for sealing many products or projects including medical packaging, specialty groceries, software boxes, electronics, toys, CD cases, sleeves, and box seals.

LINEAL PERFORATING AND/OR CUTTING CYLINDERS

Lineal cutting and perforating cylinders are multipurpose and durable precision tools which provide an optimum solution for continuous lineal slitting and perforating and for edging trimming of material webs (Figure 4.10).

Figure 4.10 - Kocher+Beck lineal perforating/cutting cylinder

They are made from high-grade materials, machined on the latest CNC machines, and provide a multipurpose tool which is a durable and high-precision product.

Lineal, or vertical, perforations are likely to be required for hand separating and are placed between rows of multiple width labels (Figure 4.11) so that the rows can be torn apart to be handled individually. In this case, it may be necessary to run some tests to determine the proper perforation requirements.

A very light vertical perforation may be provided to guide the operator in positioning slitting wheels. However, vertical perforations are not normally required in the margins as there is no need to remove the line holes from the carrier sheet.

The number of possible types of perforation is almost unlimited, with perforations suitable for all pressure sensitive papers, self-adhesive foils, composite and mono materials, cardboard packaging and metalized materials. The range is extensive including coated papers, polyethylene, polypropylene, polyamide, polyester, recycled paper, thermal paper, Tyvek and metalized material.

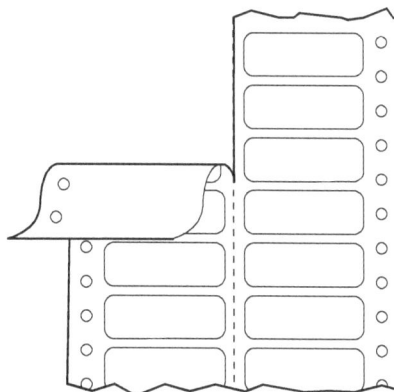

Figure 4.11 - Vertical perforation between rows

PUNCHING UNITS

Punching technology and high precision punching tools have been developed over many years and are today found in a wide range of label and related web converting applications, including:

- Smooth and serrated punching for business forms and computer labels, datamailers, tea bags, specialized self-adhesive labels and price tags, etc.
- Contour and profile punching for the corner cuttiing of ATB- and ATB2- tickets, boarding passes, wallets, parking tickets, banknotes, batteries/energizers, security labels, smart cards, solar panels, datacards, diagrams, test strips, file holes, meat tray inserts, film holes, security documents
- Surface holes for fruit and vegetable wrapping foils, band aids
- Opening features in packaging and even straw holes in liquid packaging

There are also microhole punching tools which are designed to create tear-off holes in postage stamps. The punched waste is removed through the hollow die cylinder with working widths up to 500 mm. The modules are connected mechanically or electronically to the converting machine. The module can also be installed in machines used for the production of business forms. Modification of the punch pattern or application of a security perforation can easily be achieved by changing the punch and die bars.

Holes punched for transporting purposes using sprocket or pin-fed labels (e.g. computer labels) are used to guide material in finishing lines or through dot matrix, continuous laser, thermal transfer or thermal direct printers (as seen in Figures 4.8 and 4.11). They can be realized in register to a printed web or with an equal pitch on non-printed materials. Round holes are preferred to minimize manufacturing costs as much as possible. In certain cases rectangular, oval or special shapes are used.

Although use of sprocket or pin-fed labels has declined over the years there is still a considerable residual demand for labels supplied in this format. Major printer manufactures such as OCE, Epson, Xerox and OKI still make and support printers which use this type of fan-folded sprocket stationary.

For thicker stocks such as tickets, tags, or cardstock, fanfold stationary is often preferred because it stays flat, unlike rolls which cause memory-effect curl on cardstock. Tags can be made available in various size configurations, on different stocks, and in a variety of card weights or colors for overprinting on thermal transfer printers. Finishing options include round or cut-out corners (Figure 4.12), perforating and hole punching. Tags and labels for harsh environments and outdoor use are also available. Fanfold labels, tickets and tags can be supplied in already finished, ready to use packs, or in a plain/blank or printed format ready for over-printing with fixed or variable data.

Tractor feed index cards, also known as continuous feed index cards, are used for many purposes and in many different industries, businesses, and manufacturing processes. Index cards are used as a mailing tag, mailing card, production card, pull slip, controlled substance card, procurement card,

production card, route card, inventory card, allocation card, water bills, pull ticket, transfer card, schedule card, gondola tags, strategic procurement card, maintenance card, history card, slicer ticket, student registration card, gift card, job card, notice card, transfer card, travel card, ledger card, and traffic ticket etc.

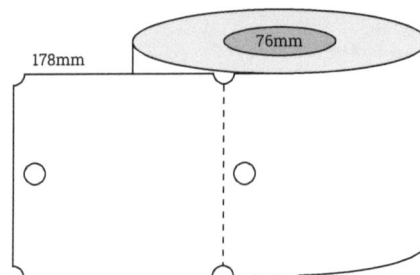

Figure 4.12 - Shows perforated and hole punched swing tags

Tractor or sprocket hole punched and perforated tickets are also to be found in use as shelf-edge perforated tickets which may be used for retail shelf labelling or pricing cards in Electronic Point of Sale (EPoS) systems (Figure 4.13).

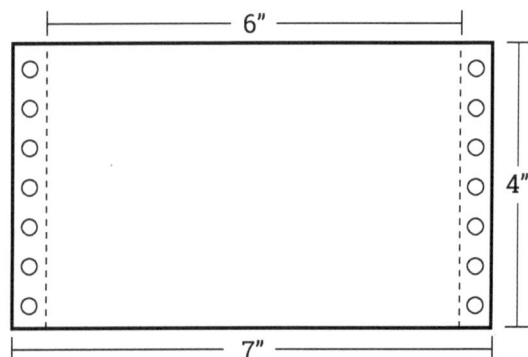

Figure 4.13 - Hole punched and perforated shelf-edge tickets

The non-adhesive blank cards are easily inserted into plastic and channelled shelf edge holders and are ideal for overprinting using laser and inkjet printing systems.

Figure 4.14 - A typical RotoMetrics punching/scoring tool

Figure 4.14 shows a typical punching/scoring tool which consists of a mandrel upon which a set of bearers and gear are fastened, plus then any combination of hole punch, score, slit, or perforation rings as required by the converter. The blades/punches are each held in place with a set screw, tightened into the 'V' groove on the mandrel. The bearer set screws are tightened into a pilot hole. The rings can be adjusted across the web and then the set screw re-tightened to the new position.

The hole cutting rings are milled and maintain standard die-cutter tolerances. As a norm, hole cuttings had been set at a distance of 1/2 inch center-to-center along the web to match the tractor feed mechanisms of the printers. The tool has evolved to include customer shapes and positioning now and can be tailored to the project as needed.

The punch, perforation or slitting rings are extremely resistant against wear and tear and can be adjusted axially, offering maximum accuracy of the perforation and slitting along the longitudinal axis. The number of rings that can be placed across the shaft is usually variable and can be adjusted to individual requirements.

These tools are generally used as under-cutters to make cuts into the liners. The feed slot or hole is cut up to the face stock, and the waste is removed by the waste matrix when stripping the face stock. Punch wheels are adjustable across the web.

SLITTING OF LABEL WEBS

Once printed and die-cut using flat, rotary or flexible cutting dies, the web of labels (2, 3, 4 or more printed labels across the web) needs to be subsequently converted into individual label widths so that each individual label can be removed from the backing liner in the subsequent label application process. Most label presses, depending on label size, converters are producing more than one label across the web width, which means that the printed and die-cut web will need slitting lengthwise at some stage into individual web widths for rewinding into the final applicator-sized reels.

Figure 4.15 - ABG International scissor knife slitting unit in operation

Some webs may also need an unwanted edge on either side of the printed web to be removed, commonly known as edge trim. Slitting operations, either in-line on the press or off-line on a slitter re-winder, are commonly undertaken in a slitting unit which utilizes cutting heads that can be of a rotary, razor or crush cut construction. Figure. 4.15 shows a typical scissor knife slitting operation on a roll-label press.

These three main slitting options can be further amplified as follows:

- **Rotary shear slitting.** This method of slitting uses male and female circular slitting knives (typically referred to as the top and bottom blades) that may or may not overlap (depending on the material being slit) and are in contact with

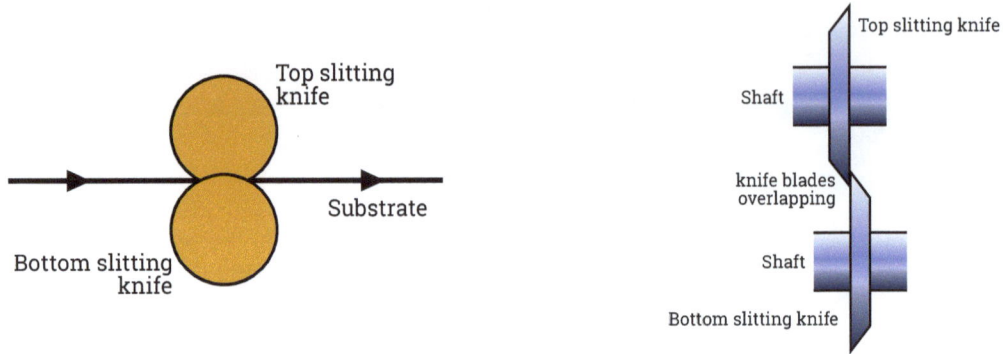

Figure 4.16 - These two illustrations show the principle of rotary shear slitting

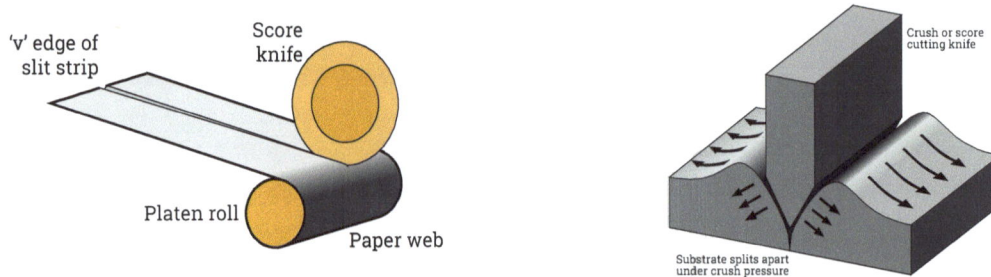

Figure 4.17 - Principle of crush cut slitting. Right shows the score knife profile and the crushing process

each other (Figure 4.16). They rotate in opposite directions to provide a scissors cutting effect. Rotary shear slitting is used to shear a wide range of materials, and similarly there are a wide range of blade designs for both the top and bottom slitter blades. The blade material also varies depending on the material to be cut, cutting application, and other factors. However, blade profile can be a key factor in overall slitting quality. Shear slitting is able to produce a good quality edge at high line speeds with little dust.

• **Crush cut slitting.** In rotary crush cut slitting a hardened steel disc cuts against a rotating, hardened cylinder, anvil roller or segment (Figure 4.17). The discs or knives that are used are dulled or 'rounded' so as to crush the substrate, rather than actually cut it. When the knife is pressed against a backing roll with sufficient pressure to crush through the substrate, then 'slitting' is achieved. For best results the web, backing roller and crush roller must all be moving at the same speed. Crush cutting, also known as score cutting, can be economical if speed and edge quality are less critical. The simplicity of score slitting is one of its main attractions, but slit-edge quality is highly dependent on important variables that must be addressed.

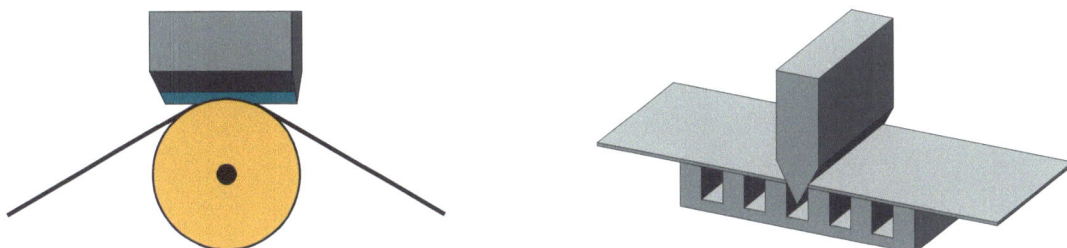

Figure 4.18 - Illustrations show the principle of a razor slitting unit and slitting process.

- **Razor slitting** is ideal for thin plastic films and is very simple and quick to set. Razor slitting units enable the fixing of sharp razor type blades which cut the substrate as it passes through the unit. See Figure 4.18. Although the razor blades are low cost, they do need to be frequently changed to ensure a good quality slit edge. Razor slitting has the lowest installed cost, being the simplest and cheapest method. It can be easily adapted to almost any machine, in almost any location and is potentially the cleanest method of slitting, assuming the appropriate materials are being slit.

Variables to analyze for each type of slitting application include the placement of slitting knives, which can be shaft mounted, mounted in individual holders on a bar, or mounted in a combination of the two. Blade materials can be anything from high chrome steel to carbide or ceramic; bevel and grind angles of the slitting blades can be varied, as can cutting angles, while the most important variable is the blade holder. Modern slitting units are capable of automatically positioning the top and bottom cutters, working from a digital program, so eliminating the need for manual setting of the slitters.

On a label press the slitting unit is normally positioned in the press prior to the press rewind unit, but may also be positioned at the unwind end of a press to ensure accurate web widths enter the press. The slitting unit will be adjacent to a waste removal or extraction unit which is used to take away the unwanted edge trim waste.

Figure 4.19 - A fully automated slitting module from ABG International

Slitter rewinders are normally used to slit printed label webs, plastic films, paper and metal foils. An unwind unit holds the roll stable and allows it to spin; it is either braked or driven to maintain accurate tension in the material. Some machines have a driven unwind which reduces the effect of inertia when starting to unwind heavy rolls or when the material is very tension-sensitive.

If the holder cannot maintain the proper settings, the entire slitting operation will be problematic. Decisions to be made before slitting are based on the process material's characteristics, such as weight, thickness, flexibility or abrasiveness. Machine working width, slit width, operational speed, the number of cuts to be made, the quality required, and even the number of set-ups per shift and the number of shifts, are also important.

It should also be noted that slitting is also undertaken on jumbo rolls of label stock that need to be slit into narrower rolls prior to printing and converting on narrow-web presses.

Chapter 5

The nature, use and manufacture of embossing dies and cylinders

The use of embossing processes to enhance luxury labels and packaging with a raised or tactile image or lettering has become increasingly popular over the years, either used on its own to provide a visual effect through the reflection of light on differing raised image surface levels on an unprinted substrate (blind embossing) or when used in combination with printing and/or foiling (register embossing). Alternatively a relieved emboss/deboss image effect (de-bossing) can be created. Other possibilities today include tint embossing, micro embossing, holographic embossing, polished embossing (glazing), sculptured 3D embossing, continuous textured areas and graining.

These various embossing/debossing processes add a distinctive elegance to many paper-based label stock or carton board substrates used for added-value decoration of wines and spirits, food and beverages, health and beauty, cosmetics, confectionery and personal care products, as well as to some quality wrapping foils, stationery, business cards or even wall coverings. Embossing/debossing is also remarkably effective when used to provide an overall textured effect and in adding impact to something which might otherwise be quite ordinary.

Embossing/debossing may also be used for more than just aesthetic appeal. It can also be used to provide practicality – particularly for the sight impaired. Raised Braille images, for example, have been required on all food products in the UK and are starting to appear on food and wine labels in the USA. Embossing can also be used to add 2D security features and holographic effects.

In the label and packaging sectors embossing tools are used to shape/set the surface of the label or card substrate to create either a raised (embossed) or recessed (de-bossed) image design, decoration, lettering or pattern that will enhance and elevate the quality and standard of the label or pack. Quite simply, embossing is used in the world of labels and packaging for a wide range of decorative and other reasons, including:

- The creation of a luxury look and feel to the product
- To provide a tactile and more pleasing surface texture
- To attract attention and draw the consumer's eye or hand to the product
- To enhance text or graphics to make a design, pattern or logo stand-out

- To add elegance and sensuality to the label or packaging
- To create a higher profile image or identity for the labeled or packaged product
- To provide a relatively inexpensive way to enhance a label or pack's look and feel
- To meet a requirement for Braille on labels or packs that may be used by the blind
- To add a more expensive or added-value identity to products
- To provide anti-counterfeiting or security features
- To add dimension or 'realism' to flat products.

These various types of embossing effects are created by applying a texture, image, text or graphics to a label or pack substrate by raising or recessing its surface at varied angles. Frequently used with hot or cold foiling, embossing or de-bossing effectively alters the surface of a label or other substrate by creating a three-dimensional raised or sunken design or image, most commonly achieved by the use of two dies: one raised (male) and one counterforce (female), which may either be flat or rotary.

image or with a foil. The color of the blind embossed image is the same as the color of the substrate surface. It can also be called a self-emboss or same color embossing. Quite simply, by creating a raised area using a die, blind embossing is able to create a subtle paper colored image that can be felt as well as seen, offering both visual and tactile appeal (Figure 5.2). It is especially effective when a subtly elegant, three dimensional image is desired. Different label materials, such as paper, film and foil, create different effects. So embossing can be a versatile option for creating a standout label.

Figure 5.2 - An example of blind embossing

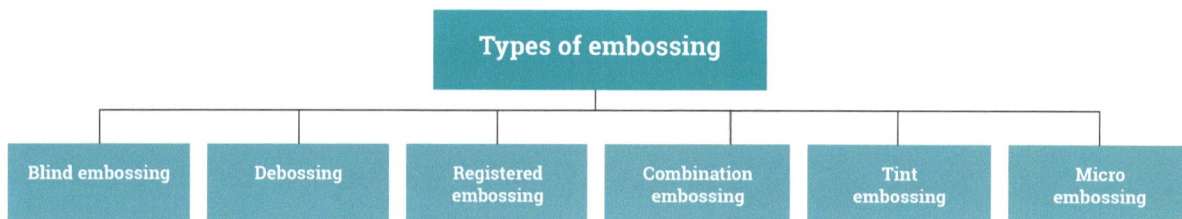

Figure 5.1 - Types of embossing used in the label and package decoration sectors

TYPES OF EMBOSSING

As already mentioned, there are different types of embossing process depending on the particular effect, image or design required and whether the embossing requires to be registered to prior letterpress, litho, flexo, screen or digital printing, or to hot or cold foiling. The main different types of embossing process are summarized in Figure 5.1 and under the subsequent sub-headings,

Blind embossing. A blind emboss is embossing which is not stamped over or registered with a printed

Debossing. In this process, the substrate surface is recessed below paper level instead of raised as in conventional embossing (Figure 5.3). De-bossing uses the same techniques as embossing to create the necessary indentation, except that the process involves the application of pressure to the face side of the substrate, forcing the material downwards into the female die so as to create the recessed profile. This can be as emphatic or delicate as the graphics or words dictate, and careful choice of substrate (avoiding bright whites and very smooth materials) will undoubtedly enhance the effect.

Figure 5.3 - A debossed image

Registered embossing. This is an embossed image that exactly registers to a prior printed image, design or text. The printed image area is embossed to give an attractive or exciting raised look, as can be seen in the detail from the award winning Dream No 7 wine label (Figure 5.4) produced by Collotype Labels North America Wines and Spirits. Embossed labels can work well in combination with four-color process printing. Depending on the specific customer requirements and specifications, the bevel can stay inside the printed image or go outside it.

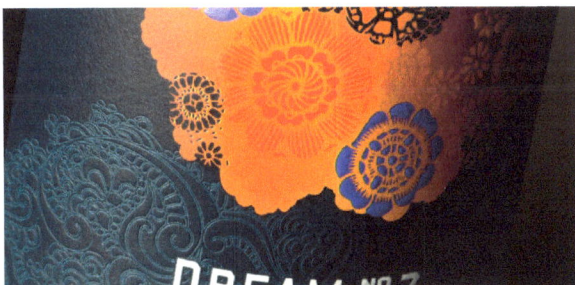

Figure 5.4 - Registered embossing on a wine label design

Combination embossing. Combination embossing in the same pass refers to the embossing of a foil stamped image and can be used to add even more interest to a gold or silver glow.

The process of combination foil stamping and embossing in the same pass with one die/tool can

only be done with single combination brass flat dies – not with cylinders. The die has a 'special edge' around the image area to ensure that the foil substrate separates the foil layers evenly from the carrier with each combination impression cycle (see Figure 5.6). An example of a substrate which has been blind embossed (right) and also combination foil embossed (left) can be seen in Figure 5.5.

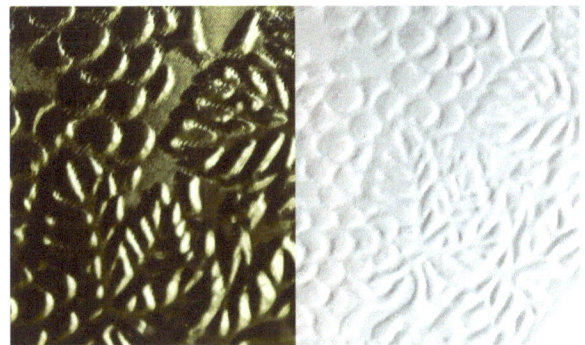

Figure 5.5 - Comparison of combination (left) and blind embossing

It is also possible to run inline one cylinder station with a foiling die and run another cylinder with an embossing die. That is two cylinders, rather than using a flat combined foiling/embossing die. This enables the press to be run at higher rotary printing and converting speeds.

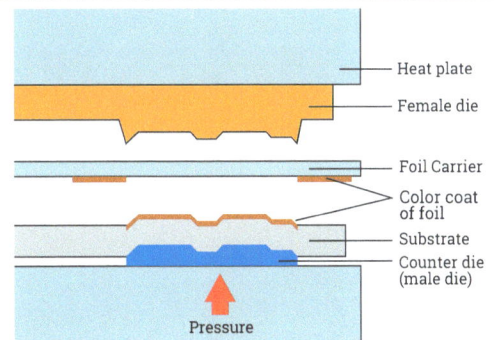

Figure 5.6 - Principle of combined embossing process

Combinations of foiling and embossing are still carried out in label decoration, but they were historically used extensively for the manufacture of 'seals' both self-adhesive and none self-adhesive applications. Seals provided a superb and unique method of product decoration particularly on high weight metalized substrates.

Tint Embossing. A relatively new method of embossing in which a tint foil is used in the embossing process. The methodology is the same as other types of embossing but the technique has become increasingly in demand. For tint embossing it is generally best to use white stock because the tint foils are translucent.

Micro embossing and debossing. This is where the effect is achieved with minimal depth but using intricate and complex designs. The process has become increasingly attractive in security applications such as event ticketing, anti-counterfeiting and legal documents. Indeed, embossing may be used for a wide range of security purposes. Embossed seals or symbols of authenticity add security features to labels, government forms, legal documents, and corporate papers. Having said that, security embossing today is perhaps considered as an older form of document or label security, as more alternatives in analogue and digital print processes, origination and pre-press, foiling, etc., continue to be developed.

Glazing. This refers to a polished emboss. It is not used in the rotary/narrow-web industry because in order to undertake glazing it is necessary to slow the dwell and 'be on impression' for a period of time. Nevertheless, glazing is a popular technique in some print sectors, particularly when used on dark colored stock. The heat and the pressure when pressing the die are increased substantially. This adds shine to the surface. If a very high temperature is used, light color papers can be scorched to change the paper color. This provides for great contrasting designs if done properly.

PREPARATION OF ARTWORK FOR EMBOSSING

As can be seen from the above summaries, there are many types of embossing to choose from depending on the embossing design created, the nature and type of substrate being used, the embossing effect required, whether the embossing is blind and raised or de-bossed and recessed, whether it is registered to prior printing or foiling.

To be able to produce the ultimate in embossing results for any of different types of embossing it is essential that everyone in the production chain understands exactly what is required and the effects desired – from the designer through to the engraver and on to the printer and converter. Clear communication, understood by all, will undoubtedly help to ensure that a quality product and finished image will be produced.

There are a number of aspects to bear in mind when preparing artwork for embossing, including:

- When creating digital files for embossing use a file format agreed with the engraver, most usually Adobe® Illustrator although others may also be accepted. Only send the tooling artwork, not the print files
- Provide artwork 100% actual size, fully stepped out and saved at 600 dpi or higher
- Artwork should be Vector art and solid line art supplied without screens or tints. Do not include masks. Grayscale artwork will not provide optimum results
- Convert all type to outlines. Avoid too many fine details and tiny criss-cross lines. Aim to keep the design as uncluttered and bold as possible. All objects should be 'filled'
- When using lettering in the design use sans serif fonts and space them out so that there is enough space between each letter to allow for the embossing effect
- The size of the artwork may need to be increased slightly to compensate for the added dimension. Check with the engraver
- Where multi-level embossing is required it may be advisable to use color coding to indicate the various levels. Again, check with the engraver
- Try to keep the image area at least .25 inches away from the edge of the web or an oversized sheet to avoid puckering or wrinkling. If the embossing is being done on a finished project, keep a .5 margin.
- The artwork should indicate:

- The gear side
- The web direction/leading edge
- The center line
- The die-cut line (if applicable).

Once the digital files have been created they should be sent to the engraver (saved to a CD) via conventional couriers, together with a completed order form for the specific engraved tool.

THE EMBOSSING PROCESS

The process of embossing is a relatively simple and cost effective means to raise or recess some areas of a paper, carton board or other surface so as to enhance its look and feel. However, there are a number of things that need to be attended to and reviewed in the creation of a successful embossing project.

These considerations include the method of embossing (flat, segmented, rotary), the embossing die materials (magnesium, steel, brass, copper, etc), the structure of the dies to be used, the die etching or engraving process, the type of surface to be embossed (paper, card, foil, etc) the creation of the necessary artwork and the particular embossing details. Each of these factors are discussed in this chapter, the choice of method generally being governed by the type of label required, the run length and the cost.

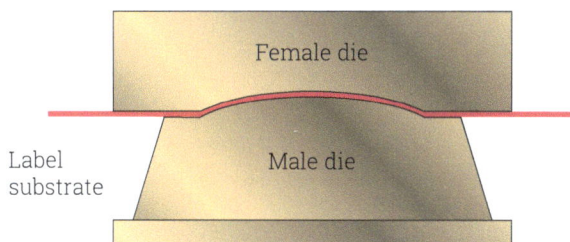

Figure 5.7 - Principle of applying pressure in the flatbed embossing process

Embossing presses or units can take many forms,

but in general the operating principle remains the same whether using flatbed, semi-rotary or full rotary. This is shown in Figure 5.7, based on a simple flatbed embossing operation. The choice of method is usually governed by the type of label required, the run length and the cost.

As can be seen, the substrate to be embossed is sandwiched between the male and female dies and pressure is applied so as to force the male relief image into the female recessed image.

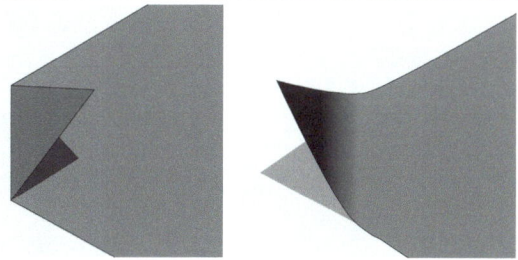

Figure 5.8 - Testing for stock suitability. Hand-fold a corner of paper stock amd check springback. If the paper holds fold 180° to 190° (left) it is suitable for rotary embossing. If it holds 90° or less, it is unsuitable.

The application of pressure pushes the label substrate into the recessed female image to create a raised profile on the label surface. The depth of the female die will govern the height of the embossed image - the deeper the die the greater the profile of the image. It is important that the substrate contacts the bottom of the female die and this can only be done effectively by pressure on the male die.

If there is insufficient pressure and the dies do not fully contact then some of the embossed image, particularly in the fine detail, may be lost. Too much pressure will damage the surface of the substrate and can in some cases puncture the substrate, destroying the embossed area.

Consideration must also be given to the type of substrate to be embossed and what depth can be achieved before the substrate fibers break. The required depth can be assessed using a simple folding test (Figure 5.8) or from the original artwork and allowing the engraver to determine the correct die depth. Synthetic materials are unsuitable for embossing as they are cannot be used to create a raised image and these substrates will not accept any direct pressure from an imaged die and are easily punctured.

Heat may be used to advantage when embossing - but it is not necessary. The use of a heated die can aid the molding of the substrate fibers during the embossing process, but care must be taken not to apply too much heat, as this will distort the substrate. However, in flat die stamping or combination stamping/embossing heat is necessary to activate the foil release layer.

No ink is used for the actual embossing process but very often the embossed image will lay over the printed area of the label. This will involve close register between the printed image and the embossed image and this is called a 'registered emboss'. If the embossed image lies in a non-printed area this is called blind embossing'.

Embossing units are available in many sizes, which are usually measured in terms of the maximum amount of pressure that can be applied.

METAL EMBOSSING AND POLYMER DIES
There are four types of metals – steel, magnesium, brass or copper – that are predominately used for the manufacture of embossing dies, with the choice depending on a number of factors such as:
- the shape and depth of relief of the image
- the texture to be created
- nature of material to be embossed
- the length of the run
- complexity of image
- depth of relief
- perceived productive life of the die
- requirements specific to product use, e.g. food or pharmaceuticals.

The flatbed and semi-rotary processes use steel, magnesium, copper and brass dies, whereas full rotary embossing generally uses only brass or steel dies. The manufacturing costs of embossing dies can be more, particular when tooling and engraving full rotary male and female brass dies. These different die materials are reviewed as follows:

Steel dies can be machined, CNC engraved, hardened and ground and are suitable for long running applications and, although they are the most expensive of the embossing die materials, can pay for themselves in terms of reduced downtime. However, they are limited with the type of engraving effects and cannot go as deep. Fine lines can also be a challenge. They can be engraved to fit contoured parts.

Magnesium dies are used for simpler embossing projects that have short runs up to 5,000 impressions on smoother stocks, with designs that are large and uncomplicated. Magnesium also allows for special hand tooling. Magnesium is the softest of the materials used for metal dies and also allows for special hand tooling. They are the least expensive of all dies. However, magnesium dies can be smashed or damage easily and must then be replaced. Best suited for flatbed use only, they are photomechanically etched. In the manufacturing process, a photo sensitive coating is first applied to the magnesium plate to be imaged. A film negative of the image to be produced is then placed in contact with the plate surface and exposed to a light source, before being photographically developed to produce the image. The plate is then chemically etched to remove the 'non-image' area leaving the 'image' area in relief.

Being a soft metal, magnesium dies may present problems when using a textured stock. They can pick-up the textured pattern from the paper stock and therefore cause damage to the die.

Brass dies are the most popular of the embossing die materials and is used for both flat and rotary embossing die production. They are very flexible and give the embosser leeway to create fine lines, sculptured images, combo foil stamping and embossing. They are also very good for images requiring extensive hand tooling and can be made by machines or by a semi-photographic process.

Flatbed brass embossing dies (Figure 5.9) are imaged using a CNC digitally driven engraving system. This method applies also to the imaging of rotary dies made from brass or steel. Flatbed brass embossing dies are engraved in the flat, with the engraving head traveling over the die moving through the X and Y axis and rising and falling as the digital data instructs. This method of engraving produces a very fine and detailed image.

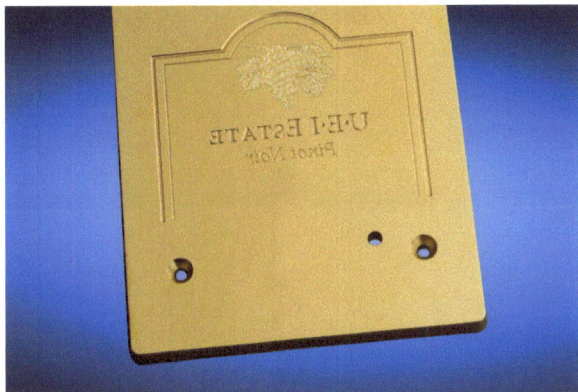

Figure 5.9 - Flatbed brass embossing die. Source UEI Falcontec

The manufacturing of the rotary embossing die requires rather more engineering than the flatbed embossing die, with the manufacturing process starting with a length of brass that is machined to the outside diameter of the required print length for the job to be printed and embossed. The ends of the cylinder are then machined to create the end assemblies, the dimensions of which are dependent on the type of press and the specification of the embossing unit being used.

Rotary embossing cylinders are made to 'very' high tolerances, with the timing of the two cylinders (in the embossing set) being critical. Set up is with the lower tool floating and the upper tool locked.

The rotary embossed image (both male and female) is imaged using exactly the same principle as flatbed engraving, but instead of the engraving head traversing on the X and Y axis, the engraving head moves only on the X axis. The rotary die then rotates back and forth on the Y axis with the engraving head rising and falling as required. This complex system of engraving is driven by a digital file which contains the image to be engraved

Brass dies are used for long runs, some in excess of 1,000,000 impressions and provide the highest

quality reproduction that wil give sharp detail and clean bevels.

Figure 5.10 - UEI Falcontec Zentastic flatbed copper die

Copper dies are used as an in between to magnesium and brass. However copper dies (Figure 5.10) do not permit hand tooling. They are mid range in price and used for runs today up to the million or so impressions, They provide better detail and reproduction of fine lined copy and images.

Polymer dies/plates. The use of male/female photopolymer plates is an alternative low price choice that is available. If the embossed detail is not very fine, and a fairly 'soft' result is acceptable, then this may be an option.

Polymer plates are made in the same manner as photopolymer printing plates but using a plate with a hard polymer, typically 55° Shore 'D' as a minimum, and with a deep relief. Offset is added to the image on either or both the male or female plates and usually a 'sprocket' pattern is added down each side of the plate to aid in registering the two plates on press. Self-adhesive tape is applied to the back of each plate, the two plates are mated to each other and then rolled, still mated together, onto the two specially made mandrels on press.

MALE AND FEMALE EMBOSSING DIES
One of the key methods of achieving embossing with rotary tooling is with a matched pair of male and female dies (Figure 5.11).

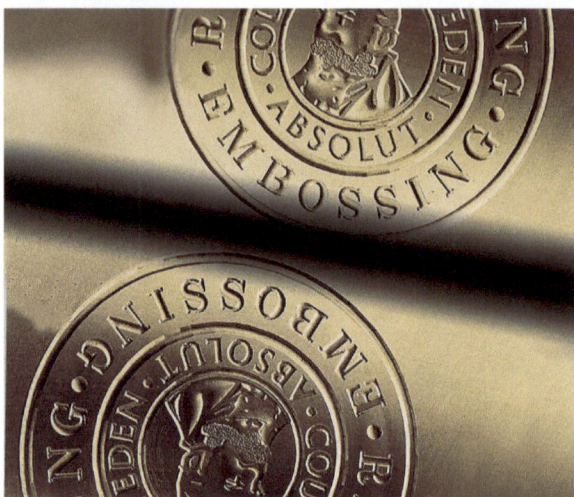

Figure 5.11 - Close-up of imaged embossing cylinders

The female die has the required image incised into the surface, whilst the male die has a matching raised image. An offset is applied to one or both images to allow the dies to accommodate the thickness of the material being embossed. This ensures that, when the dies are correctly registered, the material is 'moved' into the cavity on the female die by the raised image on the male, without crushing, cutting or unduly stressing the material being embossed.

As already mentioned, embossing dies are manufactured from a range of materials depending on requirements specific to product use, e.g. food or pharmaceuticals.

Usually steel, brass or aluminium are used, but there are a range of other materials that may be chosen based on the specific list of criteria.

REGISTRATION OF MALE AND FEMALE EMBOSSING DIES

As will have beeen evident from Figure 5.11 and the description of how embossing is achieved, there is little margin for error in the registration of male and female dies, which should always be supplied with adjustable timing gears. This enables the dies to be accurately set-up to achieve optimum results.

Of course, the gears only control registration along the web. Registration across the web has to be maintained by other means.

Two methods can be used. The first and less common is to mount the dies into a purpose made frame where cross web positioning of one die relevant to the other can be carefully set-up and maintained.

The second, and more common method, is to produce the dies with a cross-web location ring and groove. These are fitted during manufacture and lock the dies together in correct registration across the web. This is a more popular approach because it removes the requirement for the press operator to be involved in a time consuming and tedious process where errors can have fatal consequences for the dies (and possibly even the press operator). It also enables the dies to be used in non-dedicated cutter stations avoiding the need for costly investment in dedicated embossing units.

MOUNTING AND POSITIONING OF FEMALE DIES ON PRESS

There are three main systems for mounting male/female embossing die sets on a label press.

1. Die sets can be supplied as a complete set pre-registered and mounted as a self-contained embossing unit or cassette.
2. Dies are supplied without cross-web location ring and groove (but with anti-backlash and adjustable gears) for installation into a dedicated embossing unit or cassette on press. These purpose made units have accurate cross-web adjustment and clamping mechanisms to allow the dies to be correctly registered and held for production.
3. Dies are supplied with cross-web location ring and groove and adjustable gears for fitting into a suitable cutter station.

REQUIREMENTS OF REGISTER FOR MALE/FEMALE EMBOSSING.

As has already been alluded to above, registration between the male and female dies is critical and the inclusion of adjustable gears and cross-web location ring and groove assist in achieving the very close registration required with male/female die sets.

Clearance between the male and female images is, as a general rule, only a few microns more than the thickness of the material being embossed, depending on the nature of the material and the required emboss. It is evident from this that a small error in registration or small variations in the substrate, liner or adhesive, can cause an uneven emboss, cutting or crushing of the material, or even damage to the dies.

For this reason, it is also important that mounting blocks, bearings and bearing block housings are in good condition, dimensionally accurate and fit together with absolutely minimum clearance. Proofs of the embossing cylinder set should always be provided by a competent engraver.

There are die positioning systems on the market that are used to accurately place embossing dies onto a honeycomb base. These work by using a touch screen as a visual guide on where to place each die on the honeycomb chase, based off of a digital or scanned sheet. A camera moves above the honeycomb work area where the die is to be placed. This displays overlay images, one of the digital file, with the other providing the live image of the chase. As the die is moved under the camera the operator just watches the screen until the two images match. These systems are siad cut down make ready time by some 40%-70% depending on the job. The dies are placed in the correct spot the very first time.

If all of these position and registration conditions are met, and the dies are carefully and accurately set up to begin with, they will produce consistent and excellent embossing detail with virtually no wear or deterioration for the whole of the die life.

ALTERNATIVES TO MALE AND FEMALE DIES

Although male/female embossing dies have the widest application in rotary embossing, there are other approaches which, when used for the correct application, may achieve an acceptable result more economically or more conveniently.

One method is the use of a single die pressing onto a 'soft' mandrel. This avoids the need for cross-web location discs and anti-backlash and adjustable gears with the corresponding cost savings. The die can either be a male die 'embossing' from the back or 'debossing' from the top, or a female die embossing from the top. A female die is rarely used to deboss from the back.

Whether a single die and a soft mandrel can be used will depend on a number of factors:
- Sharpness of emboss required
- Depth of emboss required
- Complexity of image
- Area of material being embossed
- Type of material being embossed
- The entire thickness of the substrate – face stock, adhesive and liner – all of which play a big factor when manufacturing and embossing cylinder
- Construction and power of press
- Batch size and volume of product expected to be produced

Results from a single die can be quite acceptable, even more appropriate for some effects, than from a male/female pair but, as has been outlined above, careful consideration needs to be given to a range of factors and the choice carefully made.

FLATBED OR ROTARY EMBOSSING

Embossing has its origins with flatbed technology, but has increasingly migrated to products produced using rotary equipment and dies, with all the challenges that this different production method creates. Certainly the results achieved when embossing a particular design flatbed, may not be similar or even achievable by a rotary press running at full speed. This needs to be borne in mind when considering the transfer of a job from one method to the other. However, similar embossing effects can be achieved if the rotary press is slowed down or with the right embossing cylinder.

GRAINING

Graining is the process of converting a standard label paper into a textured finish without losing any of the durability of the stock and is a very popular technique in label production. Depending on the supplier, there are a number of different standard grain patterns available to replicate more expensive paper stocks, however it is also possible to custom make a grain pattern to suit individual company or end-use customer needs. Graining provides a more complex, three dimensional finish to the surface of paper, and under varied light conditions, highlights and shadows are revealed.

PAPER TEXTURE

As mentioned previously, paper textures play an important role in embossing. Sometimes clients select a texture paper and use embossing to smooth out the paper where it is least expected. At other times a smooth paper is used but the emboss is textured for a stunning finish.

Heavy, long fibered sheets make the best kind of paper for embossing. Lightweight, heavy coated or varnished papers are not good for embossing because they crack easily. Also recycled paper is to be avoided for embossing. In general the more processed a paper is the weaker it becomes and cannot withstand the pressures of embossing.

The depth and the degree of bevel achieved are determined by the stock. A thicker stock (including substrate face stock, liner and adhesive) can offer more dramatic embossing effects because the impression can push deeper into the paper and varying levels of relief become possible.

Chapter 6

The hot foiling process and the use and manufacture of foiling dies

The process of enhancing a product's prestige and visual attraction with a luxury metallic look, initially gold, is centuries old and traditionally was achieved by the application of gold leaf onto the surface of early manuscripts, scrolls and books by using heated hand tools. This was an expensive, highly skilled and time consuming process and was never feasible for the volume decoration of labels or printed packaging.

It was not until hot foiling or hot foil stamping, a method of heat transferring a metallic or pigment finish from a carrier strip of paper or film – known as a foil – onto a label substrate using heated dies, pressure and dwell time was first patented back in the 1890s, that the hot foiling process that we would recognise today started to become viable. This first development of the hot stamping process used both gold and coloured stamping foils made from 23 carat gold or bronze powder with a dye to obtain the necessary colour. Both these types of foils were supported on a glassine carrier strip.

The early hot stamp foiling process still had a number of limitations. The use of real gold made the foils very expensive, while bronze powder foils were easily tarnished. In addition, the coloring dyes used with the metallic powders were not light stable and tended to fade. This meant that other alternatives for metallic image printing continued to be considered.

It was not until the 1950s that vacuum or vapour metallised foils using aluminium were developed and hot foil stamping products and technology that were

to revolutionize metallic image printing throughout the whole graphic arts industry were introduced – and are still used today, particularly in the label and package printing sectors.

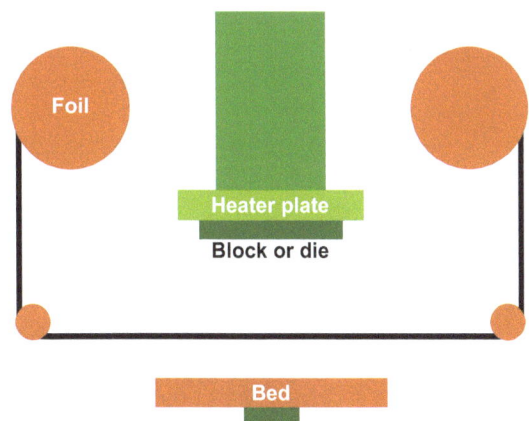

Figure 6.1 - The principle of a platen hot foil stamping machine used for label conversion

In the modern hot foil stamping process, a pigmented or metalised coated foil is transferred from a carrier, more usually today a polyester film, and fused to a substrate by heated die in a platen flatbed or rotary hot stamping machine or a hot stamping unit in a roll-label press. This means that the pigmented or metalised coating has to be compatible with the material to be stamped. For this reason, hot stamping foils are manufactured in various formulations designed to give quality prints on a specific material. Also incorporated in the foils are qualities such as abrasion resistance, oil and grease resistance, and chemical resistance.

Before looking at hot foil dies and tooling it is as well to understand a little more about the nature of hot stamping foils as these can have an impact on the hot foiling process and die manufacture.

HOT STAMPING FOILS

In principle, the structure of all stamping foils is as shown in Figure 6.2 although, depending upon the type of product involved, the release and adhesive layers are not always separate entities. As can be seen the foil is made up of a series of coatings that need to be transferred. The carrier layer is made from polyester film that can range from just under 0.0005" to about 0.0015" (12.7 microns to 38.1 microns). The thinner the carrier, the faster the foil can be expected to transfer.

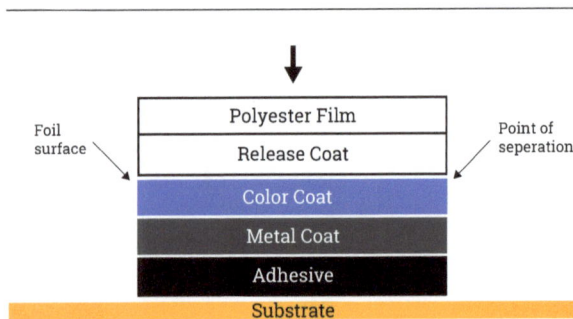

Figure 6.2 - Diagram shows the structure and composition of hot stamping foils

The first layer to be applied to the polyester film

carrier is a thin release coating (Figure 6.2), which affects whether or not a foil is suitable for fine copy or heavy coverage and, in turn, affects the cutting/separating properties of the foil. There is then a lacquer or color coat, which provides a hard abrasion resistant surface to the foil when applied and can make up to as much as 30 per cent of the total transferred layer.

The vacuum deposited metallic layer – which is around 0.05 microns – is always aluminium, so requires the lacquer/color coat to provide the various gold or other colored effects. The aluminium layer however, does not have any structural integrity. The final layer is the adhesive or sizing, which is the heat-activated layer that bonds the foil to the label or pack substrate. Thinner adhesive/sizing layers are generally better suited to smooth, high gloss materials, while a thicker layer is likely to be required on a rougher or more porous substrate.

When heat and pressure are applied to the polyester film the release coating melts and the colored metal film is transferred and adhered onto the carrier by the final layer of adhesive. This transfer and adhering process can be achieved either as a flatbed process on a platen press (as previously shown in Figure 6.1), on a sheet-fed press, or as a rotary process on rotary or semi-rotary roll fed label presses.

Over the years an enormous range of foil types has become available. Where metallics are concerned, there is every conceivable shade of gold (and silver), in bright, satin or matt finishes. These are supplemented by metallic colors, brushed finishes and a wide range of gloss and matt colored pigments, particularly white, which is used on clear materials for overprinting purposes, as signature panels and for the repair of misprinted labels. In addition, there are numerous diffraction, magnetic, holographic, patterned colored and metallic foils which are particularly suitable for security, tamper proof, brand security or background effects.

To meet more specialized requirements today, there are also fluorescent, magnetic, pearlized and fabric printing foils, together with pre-printed in-mould and heat transfer solutions. Of more recent origin is an extensive range of diffraction, patterned, embossed metallic and holographic designs, and both stock and custom-made holographic images.

THE HOT FOILING PROCESS

The hot foil stamping process can generally be referred to as encompassing three closely related processes:

- Simple hot foil stamping
- Hot foil stamping combined with embossing or deep embossing
- Security and hologram foil applications.

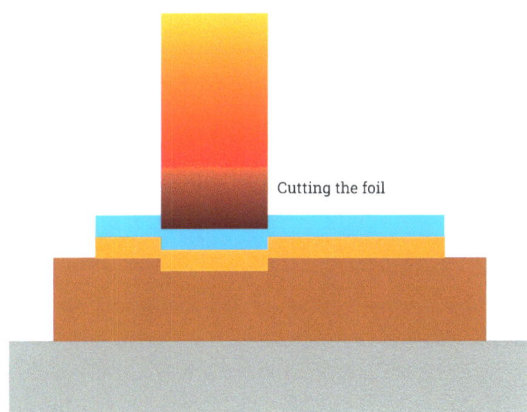

Figure 6.3 - The hot foil stamping process

In the hot stamp foiling process the colored metallic or pigmented foil is transferred to the label or packaging substrate by using a heated die to transfer the foil to the substrate under force against an anvil. To enable transfer to take place the die must be hot enough to activate the adhesive on the back of the foil and then enable the metallic or pigmented foil image to adhere permanently to the substrate.

During the transfer process the hot die has to apply enough pressure to effectively 'cut' the foil around the edge of each detail being transferred (Figure 6.3). Without this cutting process the foil will fail to transfer cleanly and may bridge across any small reverses in the copy.

Hot foiling units, whether flatbed or rotary, comprise principally of a number of key components:

- a heating system in which temperature can be

set and controlled
- the flat, segmented or rotary foiling die
- the impression platen or anvil impression cylinder, with a means of adjusting pressure.

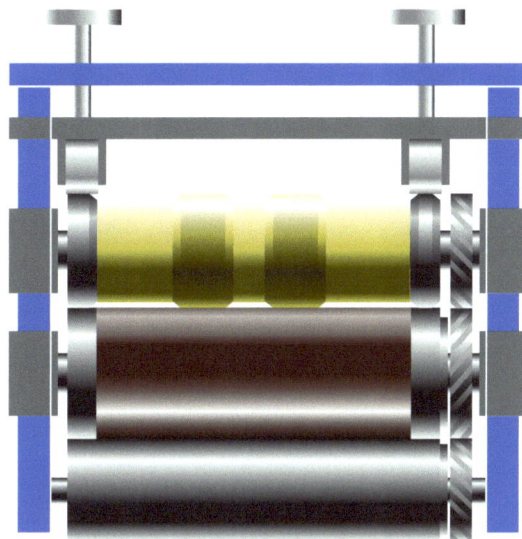

Figure 6.4 - A rotary hot foiling unit showing the heated die cylinder, anvil and assist rollers

Manufacturers of foiling units have differing approaches to the their design, operation and control, but broadly speaking it is the need to adjust and control pressure and temperature evenly across the full width of the die, and the ease with which this can be achieved, that are critical factors which will have some impact on the quality of the final product and the efficiency with which this can be achieved in production.

Hot foiling units must therefore have a means of, firstly, adjusting pressure and then maintaining this pressure at an even depth of impression throughout the foiling run. To be able to achieve high foiling speeds the latest rotary foiling units (Figure 6.4) have steel bearers added to the rotary die so that the die and anvil are sandwiched between a lower anvil roller and an assist roll which can be set and monitored. The heated die must also be able to be taken off impression when the press is stationary.

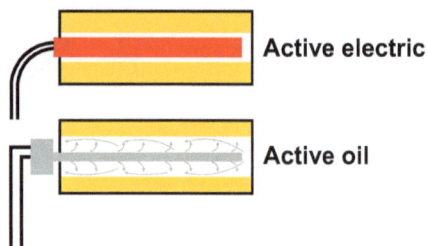

Figure 6.5 - Shows how die heating and temperature control is carried out by electrical heating or hot oil

Temperature control of hot foil stamping systems is achieved by one of two methods, of which the most common is active indirect heating via an electrically heated mandrel onto which the die is mounted and which can be used with or without cooling (Figure 6.5 top diagram). The second method is active direct heating from hot oil circulating through the internal bore of the die (Figure 6.5 bottom diagram), which, while inherently more accurate, also adds a somewhat dangerous component.

Die temperature is generally controlled by a thermostat or temperature controller which is capable of maintaining heat under normal cycle speeds – commonly in the range of 100 to 170 degrees C – and must be within a tolerance that is governed by the release factor of the foiling material and the optimum running speed of the job being foiled. This temperature must be maintained throughout the production run. Any fluctuations from the set temperature should not exceed plus or minus 5 degree C.

Temperature control only needs to be set high enough to bring the foil to a plastic state. Excessive heat will cause decomposition of the resin used in the foil and prevent its adhesion to the substrate. Other indications of excessive heat can be flaking where the foil bridges between borders (or leaves fuzzy edges which flake off when rubbed), discoloration, and dullness of the stamping.

As the die expands with heat, adjustment of the die width, diameter and image have to be made to ensure accurate registration between the foil and any ink or embossed detail. This adjustment is called the 'dispro'. For the 'dispro' to be calculated and applied correctly, the die maker will need to know what temperature the die will run at in production. To ensure a good foiled image is achieved the foiling material must leave the surface of the substrate with a clean break and some tension needs to be maintained so that the foiling material does not become loose or 'baggy'.

Obviously ease of die mounting setup, whether rotary or flat, and breakdown are also important factors, together with build quality and durability. Overall, the need for precision and quality cannot be over emphasized in the production of both the foiling unit and the die if the high quality demanded of hot foiled products is to be consistently and economically achieved.

There are three key criteria in successful hot foiling: die temperature, pressure and speed (dwell time). These criteria are interdependent and critical in achieving the demanded result on the product. How these criteria are determined and applied will depend on many factors. For example:

- What label, ticket, tag or pack substrate is being used?
- What surface is the foil being transferred to? (is it ink, lacquer, direct onto substrate etc.)
- Has the substrate surface any kind of graining, watermark or texture?
- What types of foil is being used? (metallic, pigmented, diffraction, holographic, etc.)
- The die type (flat, segmented, rotary) and the die material used (steel, brass, copper)
- What foiling equipment is being used (flatbed, semi-rotary, rotary) and what are its limitations? (max temperature, pressure and speed)
- What material is being used for the covering on the impression platen or cylinder

The inappropriate application of any or some of the above variables is the most regular cause of poor foiling results. Unfortunately, there are not any hard and fast rules, just general principles, as well as accumulated operator experience and probably some trial and error. However, to provide some basic guidelines, the general principles can be set out as follows:

- Use the lowest pressure setting possible for successful foil transfer.
- Find out from the foiling supplier what the optimum temperature is for the foil that will be used and begin by using this temperature (remembering that the die will have been 'dispro'd' for whatever temperature was provided with the die order). However, as far as the die is concerned it is usually safe to vary this temperature by up to ± 15°C without significant registration problems.
- The faster the web is traveling through the press, the higher the temperature and/or the greater the pressure that is likely to be required for successful transfer, and of course the converse is true.
- Use a hard coating with as high a resilience as possible on the impression plate or cylinder.
- Check the results regularly and adjust, in small increments, one at a time, the above variable parameters as necessary to achieve the best results.

Figure 6.6 - A modern hot foiling unit on a narrow-web combination label press. Courtesy MPS

In general, it can be expected that the transfer results will change as the web speed is raised. A slight increase in pressure will normally correct this.

While rotary foil blocking has gained ground for long-run labeling applications where other (combination press) print units may print the detail by, say, flexo or offset, the majority of hot-foiling for label printing is still short-run text and solid color, produced on small flat-bed machines.

Units designed for the hot-foil printing or decoration of labels or packaging come in a variety of configurations and widths. Stand-alone hot-foil blocking machines are generally narrow-web and operate on similar principles to those of flat-bed die-cutters. In flatbed foiling the foiling dies are either attached to a flat honeycomb chase allowing for variable positioning, or in fixed positions on a flat metal plate for repeat jobs.

Figure 6.7 - Hot foiling on Cattos scotch whisky labels printed by Royston Labels

Whichever type of die carrier base is used, the sheet and the foil are brought together in position between the die and the lower counter plate. The press then brings the upper and lower elements together to deliver uniform pressure across the entire sheet or web.

In semi-rotary hot foil stamping the flat counterplate is replaced by a rotating cylinder, with the die moving horizontally in synchronization with the cylinder rotation to deliver a narrow linear zone of pressure where the die, foil, adhesive, and cylinder all meet.

Rotary hot foiling is undertaken in a similar way to rotary printing. The dies are mounted on a cylinder that rotates in synchronization with a counter cylinder. The substrate and foil are brought together between the two in the narrow nip point.

Standard self-adhesive production widths and rotary foiling cylinders and dies are used where a combination of printing processes include foil blocking. Some hot foiling may be carried out off-line on dedicated finishing lines, such as with the majority of HP Indigo presses. Larger sheet-feed presses are mainly used for foil blocking on large sheets of glue-applied or in-mould labels, as well as in the carton printing sector.

The hot-foil process can print on a wide range of substrates and surfaces and is used to produce bright metallic effects, or to print high opacity pigmented foil colors using relatively simple equipment. The process is particularly well used to provide a luxury (metallic) look on many cosmetics, toiletries, health and beauty labels, on wines and spirit labels (Figure 6.7) and in other higher added-value label applications. The process may also be combined with embossing.

HOT FOILING DIES

Hot foiling dies are engraved metal plates, segments or cylinders where the printing image areas are in relief and raised above the non-printing areas. Whether sheet- or web-fed, sophisticated handling systems are needed to position the substrate and foil between the plate or cylinder that holds the dies and the platen or anvil that supplies the counter pressure. To produce good quality foiling dies there are a number of critical elements that are required to ensure that they perform correctly. These factors are:

- The material used for the die must be a good conductor of heat (have a high coefficient of thermal conductivity).
- The die material must be durable and have physical properties compatible with its usage and length of life.
- The die material must be suitable for the required machine engraving or chemical etching

processes.
- The image on the die must be 'dispro'd' and positioned correctly across and around the die and must be perfectly aligned to the axis.
- The die should have a high surface finish. The smoother the surface the higher the gloss of the printed foil.

HOT FOILING DIE MATERIAL OPTIONS

As already mentioned, hot foil stamping dies must be good conductors of heat and must be durable. They must be suitable for engraving and/or etching and provide for a high surface finish quality. Suitable die materials commonly used for foiling dies, depending on length of life and performance, include steel, brass, copper, magnesium and photopolymer. Each of these materials can be summarised as follows:

Steel tool material is machined, engraved, hardened and ground and is particularly suitable for long running applications and, although it is the most expensive of the die materials, can pay for itself with reduced downtime. It can be engraved to fit contoured parts.

Brass tool materials are widely used and have similar properties to steel, except that they are around a quarter less expensive than steel. They also have a better heat transfer than steel, although they are not nearly as durable. They are suitable for machine engraving and are 100 per cent recyclable. CNC-engraved brass dies are the more traditional choice for hot stamp label decoration. They are also the best choice when stamping a thick substrate or carton board stock, as the manufacturing process can accommodate such substrates.

Brass cylinders make the most efficient use of heat conductivity when compared with other current hot foil tooling options. Segmented brass hot foil dies are available that are divided into segments and/or rings. The rings are cut from brass in a similar fashion as that for solid cylinders, but each ring is limited in size specific to the image. The rings are then slid onto a mandrel, which typically is aluminum. The ring system can accommodate both electrical and oil thermal heating units.

Copper tool material is generally photoetched from artwork and has a similar durability and heat transfer to brass. Copper is claimed to be good material for ornate graphics, with the etching process being faster than mechanical engraving. A limiting factor is that coper dies can only be etched .004" deep, although open areas can be machined deeper.

For narrow-web converting there are chemically etched flexible copper dies or sleeves that are produced with a thin steel backing that is securely held to a heated magnetic cylinder. Such dies offer fast, easy and reliable changeovers, which make them a suitable choice for short- to medium-run hot foiling jobs. Copper faced steel sleeves provide a cost-effective alternative to engraved brass dies and they can be run on low cost aluminium mandrels.

Another die option enables flexible copper dies to be mounted around a mandrel using an existing OEM mechanical fastening system. This can also be an economical option for short- to medium-run applications

Magnesium tool materials are able to transfer heat as well as copper or brass, and can be photoetched like copper. Plates are etched using two chemical baths and are available in different gauges. Magnesium is the least expensive foiling die material, but is also the least durable. Different thicknesses are available.

Photopolymer dies mounted onto steel anvils are available from some suppliers, but are typically more limited to flat embossing rather than foiling due to the poor heat conducting properties of the polymer when compared to solid brass cylinders or flexible copper dies. The photopolymer, being softer than metal, will also limit the lifespan of the die.

PRODUCTION OF HOT FOILING DIES
Having selected the most appropriate foiling die material and received the required artwork specifications, the dies need to be etched or engraved (milled) as required. There are two main methods of producing the hot foiling dies: photoengraving and Computed Numerically Controlled (CNC) engraving.

Photoengraving has long been a traditional method of producing both printing plates and foiling dies. With this process the image is transferred to the surface of a plate or die that has been coated with a photosensitive substance using a photographic

negative. The background areas are then dissolved or etched away in an acid bath using a strong acid to leave a relief printing or foiling surface. Photoengraving is a relatively high overhead process, but has proved to be excellent for more complex images.

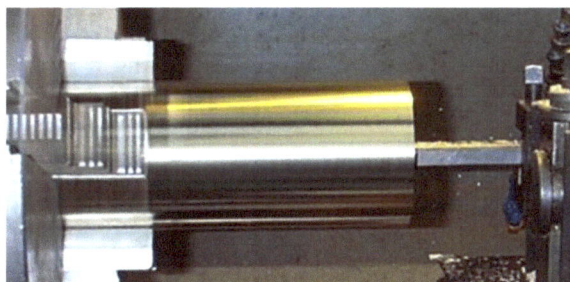

Figure 6.8 - Machining the die cylinder to correct diameter

CNC engraving is a more recent method of engraving both flatbed and rotary hot foiling dies using a computer to control a machine that does the engraving process. The image is created using specialised CAD/CAM computer programs to produce the required computer file, in turn gets fed into a particular CNC engraving machine for production. The computer software file drives the tooling heads to remove the unwanted material, enabling cylinders and dies to be produced very efficiently, particularly for less complex images.

CNC engraving machines are used to produce highly accurate engravings. CNC engraving is also great for reproducing the same, consistent results for mass production and use a wide variety of different tools – such as different drills or saws – to create a variety of products and engravings. The latest CNC engraving machines often combine many tools into a single 'cell'. In other cases a human or robotic external controller may change the parts on a machine required for engraving.

HOT-FOILING FLATBED DIES
The hot foiling dies used for flatbed hot-foiling need to be of a hard material and have a raised image similar to that used by the letterpress process. The fact that

image transfer relies upon both heat and pressure restricts plate materials to either a very hard thermoformed plastic plate for very short runs or plates produced from magnesium, brass, steel, or copper, for the longer runs.

Figure 6.9 - A flatbed foiling die

Magnesium is the softest of the materials used for flatbed metal dies and is the least expensive. The imaging of a magnesium plate is done using a chemical etching process. A photo sensitive coating is applied to the magnesium plate to be imaged and a film negative of the image to be produced is then placed in contact with the plate surface and exposed to a light source before being photographically developed to produce the image. The plate is then chemically etched to remove the 'non-image' area leaving the 'image' area in relief.

Copper dies are imaged using an etching process similar to magnesium die etching. The copper die is harder than magnesium and therefore more suitable for longer production runs and multiple image work. Good image etching characteristics will give excellent foiling results.

While magnesium and copper dies are chemically etched, flatbed brass dies are imaged using a CNC (computer numerical control) digitally driven engraving system. This method of imaging applies to both flatbed and rotary dies. Flatbed dies are engraved in the flat. The quick separation of the foiling material from the substrate surface enables very fine detail to be achieved.

Figure 6.10 - Kocher+Beck magnetic flat base

The metal dies used in the flatbed foiling process are not as sophisticated as those used in the full rotary foiling system. The flatbed die does not have any curvature issues and can be etched or engraved in the flat. Once the image has been engraved or etched the die is ready for mounting in the press.

In flatbed foiling the foiling dies are either mechanically or magnetically attached to a flat magnetic or honeycomb base allowing for variable positioning, or in fixed positions on a flat metal plate for repeat jobs. Figure 6.10 shows an example of a magnetic flat base.

SEGMENTED DIES

Segmented dies provide a means of mounting individual brass dies onto a full rotary honeycomb cylinder without the need to use an expensive solid brass die. The curved segment dies are placed in position on a honeycomb cylinder configuration and held in position using clips located along the edges of each individual die.

Once they have been located and placed in their correct positions, each individual curved segment die is securely fixed to the honeycomb base using a special tensioning key to fully tighten the securing clips. For multi versions or variations of labels it is easy to remove and replace one or more die segments as required, which can be particularly cost-effective.

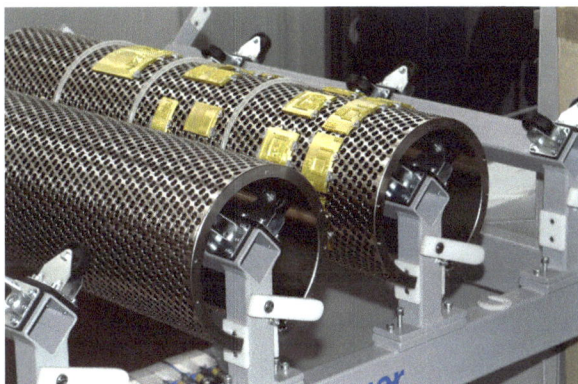

Figure 6.11 - Segmented brass foiling dies mounted on a rotary honeycomb base

ROTARY HOT FOILING DIES

Full rotary hot foil stamping dies are metal cylinders, usually produced from brass or copper alloy, although other metals can be used. The metal must be able to hold as much thermal energy on the surface as possible. Usually machined out to be hollow on the inside so that they can be heated, the cylinders are CNC engraved with the desired image on the outside. It is this heated engraved rotary die that transfers the metalized or pigmented coating from the foil carrier onto the substrate.

Solid rotary dies are imaged using exactly the same principle as flatbed engraving, but instead of the engraving head traversing on the X and Y axis, the engraving head moves only on the X axis. The rotary die rotates back and forth on the Y axis with the engraving head rising and falling as required. This complex system of engraving is driven by a digital file which contains the image to be engraved. Figure 6.12 shows a finished rotary die with engraved images.

Today, CNC engraving can reproduce even the smallest text and artwork with precision, all engraved to the proper depth on the cylinders to allow for perfect foil transfer. Polishing of the die may also be undertaken to refine the cylinders even further and to provide a mirror finish, resulting in brilliant foil transfer.

The impression cylinder, die or roller, is the support for the rotary foiling die and works in tandem with the die to press the foil onto the substrate. This support surface should be blemish free and have an ability to 'give' a little when in direct contact, through the substrate and foil, with the image on the die. But any compression of the surface of the impression cylinder must be fleeting, returning to its uncompressed state almost instantly. This 'resilience' is vital when choosing a material for coating the cylinder. In addition to its resilience, the material must be able to withstand temperatures up to 250°C and have a hardness in the order of 90° Shore.

Figure 6.12 - Kocher+Beck full rotary hot foil stamping cylinder

SLEEVED ROTARY FOILING DIES

Another option for creating a rotary foiling die involves a sleeve system. This type of cost effective rotary hot foiling system reduces the metal content of engraved rotary dies by chemically etching a sleeve that is manufactured to fit the width of the specific image area required. This can be seen in Figure 6.13 in which, in this case, a chemically etched copper faced steel sleeve is produced that can be mounted onto a low cost aluminium mandrel.

The sleeve is slid onto the mandrel and can be positioned both laterally and circumferentially to give the correct registration position. The mandrels are considerably lighter than brass or magnetic foiling dies and are easier to handle and set up. End adaptors in mandrels for oil heated foiling units remain in place at all times reducing set up time for press operators.

Figure 6.13 - Shows an Econofoil/UniSleeve hot foiling system manufactured by U.E.I Falcontec

Foiling sleeves are lightweight, easy to handle and very economic to transport with customers befitting from reduced freight costs.

Figure 6.14 - Magnetic base cylinder used with UniFlex flexible foiling dies. Source U.E.I Falcontec

MAGNETIC FLEXIBLE FOILING DIES

Magnetic foiling dies work on the same basic principle as wraparound flexible cutting dies used for the profile die-cutting of the label shape. The foiling die is a chemically etched flexible plate or 'shim' which has a steel backing and a copper surface layer which carries the foiling image. Copper is used because of its good heat conductivity. The imaged shim is positioned onto a heated magnetic cylinder (see Figure 6.14) and the cylinder and die are mounted in the foiling unit in the same way as a solid rotary die.

Imaging of this type of flexible foiling die is done using the same type of etching process as that used for imaging of copper or magnesium flatbed dies.

Fitting and removing of the flexible foiling dies is a simple operation particularly as some magnetic cylinders are fitted with pins that correspond with locating holes in the shim itself, making registration an easy operation. Flexible foiling dies are easy to changeover between production jobs.

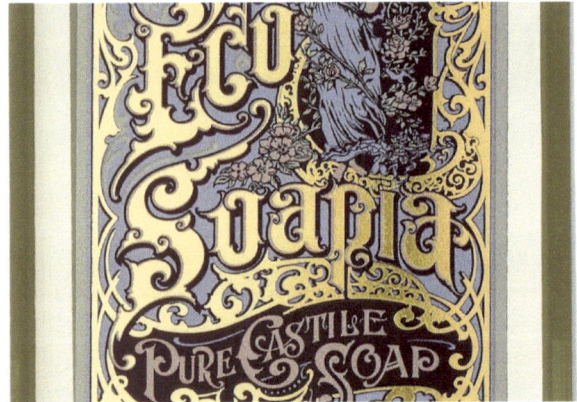

Figure 6.15 - An example of cold foiling produced by Royston Labels for Ecosoapia

COLD-FOILING

A more recent development of hot-foil blocking is the cold foil process, in which a print unit is used to print a special adhesive on the label web where the metallic effect is required. When the metallic foil is brought into contact with the adhesive it adheres to it to produce the printed foil design on the label. Cold foiling provides another option for product decoration, offering higher running speeds when compared to hot foiling.

Cold foiling will normally give the best results on

smooth and high gloss substrates. Semi-gloss papers will require a coating to be applied if the best results are to be achieved.

Figure 6.16 - Illustration shows an example of the cold foiling process

The idea behind cold foiling is quite simple: using a photopolymer plate, an image is printed onto the label substrate with the use of special UV-curable cold foil adhesive. UV light from a UV curing unit then activates the adhesive. The foil material is brought into contact with and bonded to the adhesive and substrate, with the non-bonded waste being removed. The extracted foil that is affixed to the printed adhesive is what creates the cold foiled image. See Figures 6.15 and 6.16.

While there is a place in product decoration for cold foiling, the overall quality and consistency is still regarded by many as not quite as high as that of hot stamping. For high-end label applications, which is largely the case in the health, beauty, wine and spirits sectors, hot foil remains the preferred process.

Another benefit of the hot foiling process is that it smooths the fibers of the substrate, so providing an ideal base for ensuring the highest sheen and brilliance the foil has to offer. In cold foiling, the foil is laid on top of the adhesive and fibers of the paper, which can cause the resulting foil image to be less brilliant/shiny in appearance when compared to hot stamped images.

Cold foiling also has an added expensive component – the adhesive. This additional cost needs to be balanced against the desired effect when choosing the type of foiling that is best for any particular end-use application.

MAKING FOIL STAMPING WORK

The use of hot, and cold, foil decoration has become an increasing part of the label converter's business as brand owners seek to distinguish their products from competitors. Rotary hot foil stamping in particular has continued to be a popular choice for all types of foil-decorated labels, and if all the necessary foiling parameters are correct, then the result is a clean, crisp and consistent foil transfer, making foiling an important production process amongst competitive label printers. However, it is key that the converter ensures good quality foil transfer at maximum production running speeds. Attention therefore needs to be paid to the following:

Foil selection. Foil manufacturers today have an array of foil selections to maximize the performance of the job being produced. Foils are available for porous stocks, plastic stocks, fine details, large solids and over-printable foils, so confer with the foil supplier to make the correct selection. Keep the foil rolls properly labeled as it is virtually impossible to identify the foil products without a label, especially when they are in similar color shades.

Testing of label stock. Most label stocks today are able to accept hot foil stamped images. However, there are some that may require a topcoat or primer in order to overcome porosity and to give the stock a better base for adhesion of the foil. It is therefore recommended that the stock to be used is tested with the selected foil to see if a coating or primer is required. With clear plastic substrates being a popular choice for adding decorated foil there will be a requirement to select a low temperature foil to keep the material from wrinkling or melting.

Choice of foiling anvil. While most hot foiling jobs can be run successfully with a 90 durometer Shore compound, a harder 100 durometer anvil may be a better choice for use with textured stock so that the tool can compress the substrate texture without embedding into the anvil. Compounds must be able to withstand 400 degrees F and be resilient.

Use of magnetic cylinders. Magnetic cylinders can be used with a flexible foiling die or with engraved brass dies. Engraved brass dies are still recommended for very long runs and/or jobs that continuously will repeat. However, for shorter runs, a flexible die with a metal back can be extremely effective and much less costly for the customer.

Getting started. Ensure that proper training is included with any new installation. Inline hot foil stamping adds a completely new capability to a label press and will require on-press training time to fully master for the best results. However, with the proper pre-planning, acquired technical knowledge and the proper training, converters can be up and running quality foil stamped labels in a matter of days.

Chapter 7

Cylinders, anvils, support rollers and magnetic cylinders

Much of this book has been about high performance and extremely accurate tooling used for cutting, perforating, hole punching, hot stamping and embossing using flat, solid, segmented or flexible dies. Cutting dies have to be machined and precisely sharpened to very tight tolerances so that they only cut through the layers of material needed. Male and female embossing dies need to mate exactly together to form the necessary embossed or debossed surface structure. Foiling dies also need to be very precise in structure and format to provide clean separation of the foil.

Quite simply, no label or package printing tooling can be manufactured without tight and specified tolerances. Even then, there may be variables in these processes that can have an impact on the final result. This may include set-up, bearers, pressure, distortion, substructure, temperature, materials used, etc. It is important to minimize as many variables in the whole process as possible.

Key to the performance of much of the label printing and converting process is the manufacture, quality and operation of the many different types of cylinders, anvils, sleeves and rollers that are used, together with all the associated bearers, shafts, gears and frames. This chapter examines some of the main cylinder types used in the industry in rather more detail, as can be seen in the following flow chart.

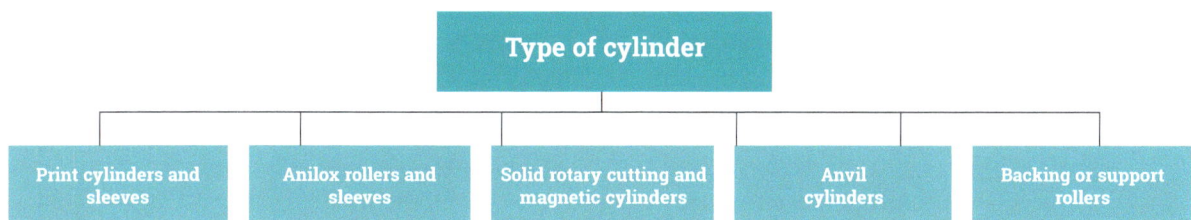

```
                        ┌─────────────────────┐
                        │   Type of cylinder   │
                        └─────────────────────┘
   ┌──────────────┬──────────────┼──────────────┬──────────────┐
┌────────┐  ┌────────┐  ┌────────────┐  ┌────────┐  ┌────────────┐
│ Print  │  │ Anilox │  │ Solid      │  │ Anvil  │  │ Backing or │
│cylinders│  │rollers │  │ rotary     │  │cylinders│  │ support    │
│and     │  │and     │  │cutting and │  │        │  │ rollers    │
│sleeves │  │sleeves │  │magnetic    │  │        │  │            │
│        │  │        │  │cylinders   │  │        │  │            │
└────────┘  └────────┘  └────────────┘  └────────┘  └────────────┘
```

Figure 7.1 - Types of cylinders, sleeves and rollers used in label printing and converting

In flexographic, offset litho, gravure, semi-rotary and rotary letterpress printing, the term 'cylinder' normally refers to the rollers on which the printing plates are mounted (plate or print cylinders). In offset litho there is additionally a cylinder that carries a blanket (blanket cylinder) onto which the plate prints and then transfers or 'offsets' the printed image onto the printing substrate. As can be seen in the Figure 7.1 diagram, in addition to print cylinders, other examples of cylinders and sleeves used in the labeling industry are: print sleeves, anilox rollers or sleeves, solid rotary cutting cylinders, magnetic cylinders, anvil cylinders, backing rollers or support rollers.

Before going on to examine these different types of cylinders and rollers in rather more detail, it would be useful to look at the various elements that are common to many of them and their relative positions in relation to each other. Figure 7.2. shows a typical die, magnetic, or support cylinder and its key components.

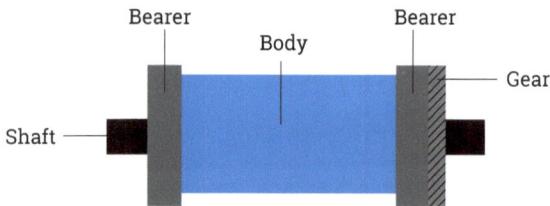

Figure 7.2 - Components of a typical cylinder

Bearers are the areas that provide a smooth rotary movement for cylinders that come into contact with each other during printing and die-cutting.

They are typically placed on outer edges of printing or cutting areas. During the printing process, they can minimize plate bounce and over impression by helping to carry part of the impression pressure or load. In the die-cutting process, they support the cutter against the anvil and provide sufficient space between the peak (top) of the cutting edge and the surface of the anvil to ensure adequate penetration of the cutter.

Where bearings are part of the design, they can

be subject to wear and it is advisable that replaceable parts are held available to avoid any disruption in the event of a breakdown. Press and converting units are subject to stresses and strains caused by the use of heavy tooling, continuous running and potentially inadequate maintenance. Cylinder bearings have a limited life so regular maintenance should be in place, including cleaning, checks for wear and adequate lubrication.

Shaft. Printing and cutting cylinders may have the shaft as a permanent part of the cylinder body, or have a removable shaft, bearers and gears. Where shafts are removable there will normally be a key and keyway (Figure 7.3.) to connect the cylinder various parts together. This whole system is called a keyed point. This keyed point still allows relative axial movement between the parts.

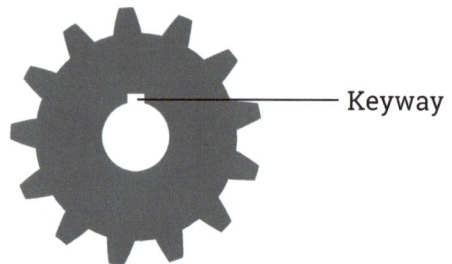

Figure 7.3 - A sprocket/gear with an internal parallel keyway

A shaftless drive is one in which the mechanical drive shaft has been replaced by an electronic drive shaft and where single driven (servo) motors are in use.

Gears. When gears are manufactured it is intended that they should be meshed at a certain depth with the gear they are driving. When meshing the printing or cutting cylinder gear with the adjacent roll gear, the diameter is critical to accurate registration. Too deep or too shallow a mesh in the print station may cause loss of register between one color and the next. Regular lubrication of gears is preferred to intermittent applications, as this will maintain a constant film of oil and even out temperature fluctuations.

Cylinders are made using mainly an aluminum material but steel is possible. They are supplied with the standard gears in high accuracy or with hardened and ground gears. There are treatments available to protect various cylinder surfaces. As the norm, cylinders are supplied with the gears and may include additional components such as shafts, bearers, register rings and other elements as required.

CYLINDER AND ROLLER PRODUCTION

Cylinders used in the label printing and converting process are mostly machined from solid materials on CNC milling machines. Milling machines are often classed in two basic forms, horizontal and vertical, referring to the orientation of the main spindle. Both types range in size from small, bench-mounted devices to room-sized machines. Unlike a drill press, which holds the solid tool stationary as the drill moves axially to penetrate the material, milling machines also move the solid tool radial against the rotating milling cutter, which cuts on its sides as well as its tip. Solid tool and cutter movement are precisely controlled to less than 0.0001" / 0.00254 mm).

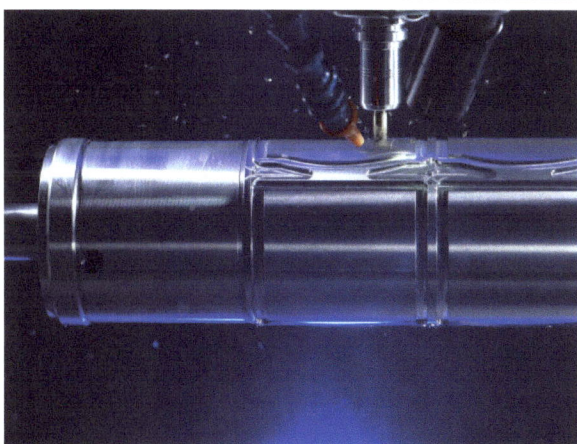

Figure 7.4 - CNC milling of a cylinder. Source: RotoMetrics

CNC milling machines can perform a vast number of operations, from simple (e.g., slot and keyway cutting, planing, drilling) to complex (e.g., cutting

lines, embossing patterns, etc.). Cutting fluid is often pumped to the cutting site to cool and lubricate the cut and to wash away the resulting swarf (See Figure 7.4).

There are a wide variety of milling machines available, including multi-axis machines in which CNC controlled tools move in four or more ways, manufacturing parts out of metal by milling away excess material. Multi-axis machines also support rotations around one or multiple axes.

Final finishing operations on milled cylinders are likely to include processes such as cylindrical grinding and polishing.

PRINTING CYLINDERS (PLATE CYLINDERS)/ PRINTING SLEEVES

Printing cylinders. Standard printing cylinders form the basis of every label printing machine and are manufactured with the greatest care in order to guarantee optimum fit and run-out accuracy.

Figure 7.5 - Printing cylinder with bearers and gear wheel. Source: RotoMetrics

In the flexographic, letterpress and litho processes, the printing plates are located on the print cylinders. Each cylinder needs to have accurate and even contact with the inking rollers and the surface of the substrate. In the case of the litho process, accurate and even contact is with the offset blanket. All print cylinders should run perfectly true with an accuracy of .0001" / .0025 mm ensuring that the pressure on the adjacent rollers is consistent. The diameter of the printing cylinder must be corrected to compensate for the thickness of the printing plate and mounting tape.

Printing cylinders (Figure 7.5) used in the roll-label industry include plate cylinders, blanket cylinders and impression cylinders. These are made from solid aluminium or steel, or produced as a tube with end rings fitted. Cylinders may have spur or helical gears, no less than AGMA Class 9 for precision and quAdded-value protection can be provided with various coatings for the surface of the cylinder. For print cylinders, anodizing or other hard surface treatments can prevent scratches or damage during demounting of the printing plates. If required, they can be provided with a range of non-corrosive hard wearing surfaces.

For cylinders with bearings, the bearing quality is critical for fine process printing. When ordering plate cylinders, convertors should be aware that not all bearings of the same size or product code are of the same quality. Choose a supplier who has tested and proven the quality of the bearers that they utilize.

Printing cylinders may be refurbished and treated/ re-treated with a range of hard wearing and non-corrosive surfaces to suite particular applications. During refurbishing, shafts and journals are checked and closely inspected for fatigue or wear and refurbished or replaced to restore them to their original condition.

Figure 7.6 - Shows an example of a rubber coated cylinders. Source: Rotometal Company

Coated cylinders. There are a number of suppliers that offer unique coatings that can be applied directly onto existing or newly manufactured cylinders. The coatings provide a platform for a plate that will enable consistent impression, so eliminating a huge variable of the flexo process. See Figure 7.6 for an example of a coated cylinder.

Printing plates are mounted directly to the standard cylinder with double sided tape. The most commonly used tape thickness is .015" / 0.381 mm but others are available.

Rubber coated cylinders are also available as varnish, lacquer and lamination cylinders. The type of rubber and its degree of hardness ordered according to customer request.

Printing cylinder sleeves. A conventional printing cylinder normally consists of the cylinder shaft with bearing system, a gear and the printing cylinder body. Such cylinders, depending on the printing length, may be quite heavy, somewhat awkward to handle, and may require considerable maintenance. Special bearing options are needed and great care needs to be taken when inserting such cylinders in the press.

Figure 7.7 - On the left, a conventional printing cylinder with gear and bearer rings; on the right, a printing cylinder sleeve

Other designs may include a lightweight print cylinder sleeve made up of one single component. Its bearing system is straightforward and extremely easy to insert into the press. In addition to this easier handling, printing cylinder sleeves also save a great deal of time during setup and immediately after the print job. The printing cylinder sleeves are designed to move into the sleeve changing position at the touch of a button, and the change itself will take just a few moments. While the new printing cylinder sleeve is still moving into position, a direct servo drive will be presetting the print registration.

A particular advantage of printing cylinder sleeves is that they do not have gears (see Figure 7.7), thereby eliminating the gear marks that tend to appear in the print image over time. Gear wear may often be the culprit when experiencing gear marks in the print.

Printing cylinder sleeves that do not have gears have a significant advantage during printing. Over time, changing gear combinations – such as those between printing and impression cylinders can result in gear marks (called 'barring') appearing in the print image. With conventional printing cylinders this barring may have a variety of causes but one could be different degrees of wear or even very slight damage to the teeth of the printing or impression cylinder gears. This can be explained using Figure 7.8 below.

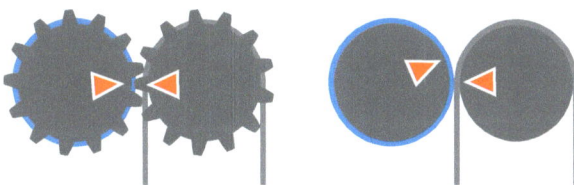

Figure 7.8 - Comparison between geared printing cylinders and servo driven cylinder sleeves

The diagram on the left shows a conventional printing cylinder with gears. This print cylinder cannot perform independent register corrections. There are other designs, however, that adjust the geared anilox to the geared print cylinder for registration moves with no impact on the web tension. The printing cylinder sleeve on the right, though, is able to perform register corrections without the impression cylinder (no impact on the web tension).

The printing cylinder sleeve makes systematic use of the benefits offered by a press's state-of-the-art servo drive concept. In conjunction with a chambered doctor blade and anilox roller sleeve, this concept enables extremely quick job changes and highly reliable processing of a wide range of substrates.

Originally developed for mid/wide-web flexographic printing, printing cylinder sleeve systems are also now used in narrow-web label printing. Making all this possible was the introduction of servo technology, paving the way for the industrial use of synthetic printing cylinder sleeves, with a number of suppliers offering different sleeve constructions.

Synthetic printing cylinder sleeves are considerably lighter, can be manufactured at competitive prices, and make larger printing cylinders much easier to handle. Composite sleeves with hard inner and outer shells made of plastic have been particularly successful in establishing themselves in the market. In most cases, the sleeve itself mostly consists of a two-component plastic foam.

These systems are very light and can be manufactured at competitive prices, but they tend to age and are not necessarily resistant to solvents or cleaning agents. Over time, the printing cylinder sleeve thus loses its original dimensional accuracy having a significant impact on print quality. Depending on printing requirements and frequency of use, the service life of a synthetic printing cylinder sleeves can be anything from just a few months to 2-3 years. After this, the sleeve has to be replaced. Alternate designs are available that incorporate an inner synthetic sleeve and an outer aluminum shell. This design can greatly improve the life of the sleeve cylinder making it more comparable to the life of a hard-coated aluminum print cylinder.

Aluminum printing cylinder sleeves are only marginally heavier than their synthetic counterparts, but are extremely precise and robust. They are fitted onto a high-precision printing cylinder shaft and held in place by a clamping system at the touch of a button. Unlike synthetic printing cylinder sleeves, the aluminum sleeve does not age and retains its dimensional accuracy over its entire service life. What's more, with good care and handling, there is very little wear-related ageing or maintenance with aluminum sleeves, making them a potential one-off investment.

ANVIL CYLINDERS/ROLLERS

Anvil cylinders are hardened steel rollers upon which the bearers of a rotary die, magnetic cylinder or perforation cylinder run. Normally, this cylinder is placed in the bottom position of a die station. However, for certain jobs it is necessary to place the anvil roller in the top and the cutting cylinder in the bottom position. Most often there is a roller under the anvil, called a support roll.

Figure 7.9 - A typical anvil cylinder. Source: Kocher+Beck

Anvil cylinders (Figure 7.9) are characterized by their exemplary hardness and run-out accuracy, whether in use as a standard diameter cylinder or with a plus or minus body diameter (called a 'stepped' anvil). Together, these anvils can compensate for a wide range of backing materials. They are quality tools that guarantee success. They are available for all machine types, in all standard designs and, of course, special designs are also possible (based on drawings or samples). Straight anvils are used for most standard daily operations, and stepped anvils provide the flexibility to run dies on liners other than those for which they were made. For some materials, most often paper, the stepped anvil can be used to compensate for worn cutting tools, extending the life of a die when most needed.

The hardness and the finishing (roughness) of an anvil roller will normally be specified in the drawing of a machine-supplier. The base materials used for anvil cylinders are high quality vacuum hardened tool steel or case-hardened steel. The manufacturing process involves sawing/turning the steel bar, milling of the keyway, cutting the gears and vacuum or case hardening the cylinder, case hardening the gear, cylindrical grinding and polishing of the cylinder.

SUPPORT ROLLERS

These are cylinders/rollers which are positioned below the anvil cylinder in the cutting unit of a roll-label press. They support the anvil and cutting cylinder, relieving the pressure and weight of the tools off the journals of the anvil (Figure 7. 10). Having a support roll can decrease problems with flex (deflection) in the cutting cylinder when increased pressure is applied.

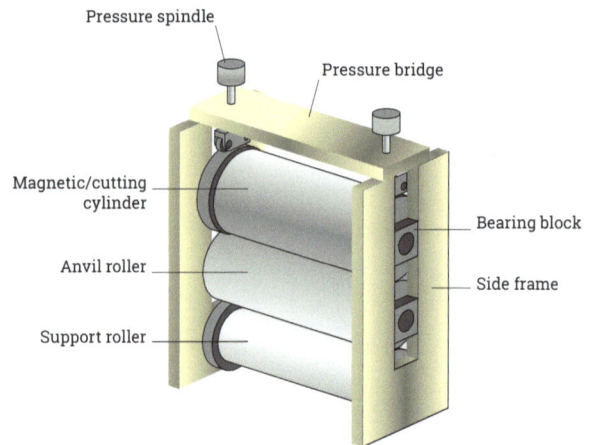

Figure 7.10 - Shows the relative positions of anvil, support, and magnetic cylinders/rollers in a cutting unit

MAGNETIC CYLINDERS

Magnetic cylinders used with flexible dies provide an economic alternative to standard rotary die-cutting tools. They are manufactured on CNC machines from high tensile and high alloyed stainless tool steel with fully hardened bearers. Hard ferrite or ceramic and rare earth permanent magnets, hardened bearings seats as well as bearing necks with fully hardened centering sleeves are usually standard. Higher-strength magnet configurations are available based on the application.

Magnetic cylinders are precision-engineered to optimize flexible die accuracy. The combination of flexible dies with permanent high-adhesion magnetic cylinders (Figure 7.11) ensures accuracy even in the most challenging applications, allowing converters the opportunity to increase profitability. Magnetic cylinders are available for a full range of label presses and converting machinery, allowing converters to use flexible dies in many different applications.

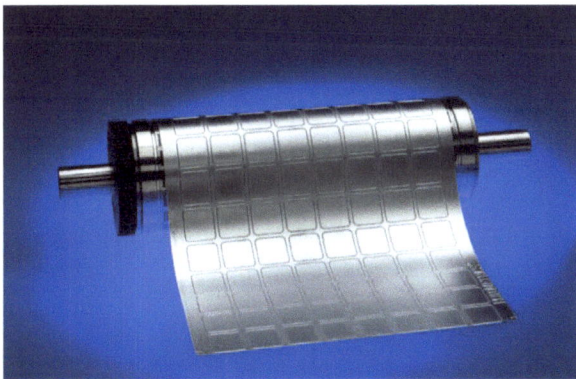

Figure 7.11 - Magnetic cylinder and flexible die. Source: RotoMetrics

Excellent cutting results not only require an accurate flexible die. A true-to-size magnetic cylinder is equally vital. Ideally, to optimize performance and life, the circumference of the magnetic cylinder as well as the anvil cylinder is recommended to reach at least the maximum working width of the press. Basic requirements for label production on thin PET liners are magnetic and anvil cylinders with .0001" / 3μm run out accuracy.

The bearings of the magnetic cylinders, the anvil cylinder and their substructure, as well as the general stability and stiffness of the cutting unit are all things to take account of to achieve accurate die-cutting, especially with thin liners.

Suppliers of magnetic cylinders are able to offer a range of services to refurbish and repair cylinders, including magnet repairs, bearing replacement, gear replacement and cylinder grind-overs. Anvils can also be reground to within microns and generally be brought back to specifications.

ANILOX ROLLERS

While anilox rollers are not a specific feature of this particular handbook it was still decided to make some mention of them while discussing the different types of cylinders and rollers used in roll-label production.

Anilox rollers are engraved or ceramic-coated rolls used to meter ink to the raised (image) areas on the relief printing plate used in the flexographic inking system. Each type of flexographic press uses an anilox roll, the surface of which is engraved (Figure 7.12) at one of three angles with a pattern of tiny cells of fixed size and depth that transfer the ink to the plate. The cells are so small that they can only be seen under magnification. The size and number of these cells determine how much ink will be delivered to the image areas of the plate, and ultimately to the substrate. An anilox roll today is either copper engraved and chrome-plated, or ceramic-coated steel with a laser-engraved cell surface.

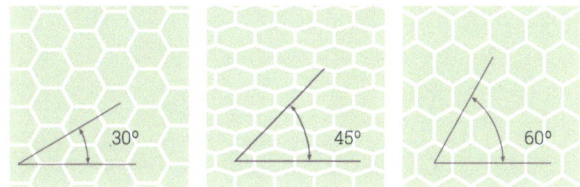

Figure 7.12 - Anilox cells are engraved at one of three angles: 30°, 45° or 60°

There is also some use today of anilox sleeves. These are not new. They have been under development, testing, trials and use for a number of years, with many manufacturers of 'gearless' presses particularly looking at the benefits of sleeve technology, such as ease of register, overall quality, low weight, maintenance, etc, as well as lower shipping costs and storage capabilities.

Although early attempts at anilox sleeve manufacture were somewhat hit-and-miss due to a variety of reasons there have subsequently been dramatic improvements in the material, construction and stability of anilox sleeves.

In use, anilox sleeves (Figure 7.13) need to be expanded to fit securely on the press mandrels. This may be undertaken with a mechanical mandrel that expands by hydraulic action, or through the use of a press mandrel with an inner compressible layer, or bladder, that is activated by air.

Figure 7.13 - An anilox sleeve. Source: Harper

Mandrels used with anilox sleeves are critical to the success of the process, with diameter, circularity and concentricity all important.

It is usual to specify anilox rollers or sleeves by a line screen count: the number of cells in a given linear distance. The higher the line screen count the greater the number of cells, and the smaller the cell size. The smaller the cells, the smaller the volume of ink carried in the cells. The depth and shape of the cells is also a factor in the efficient delivery of a uniform ink film to the printing plate. Cell depth ratios are critical in choosing the right anilox roller for a particular application. To avoid anilox moiré, film or plate screen angles should be at least 7.5° away from the anilox cell angles.

The line screen count of anilox rollers and sleeves – expressed as lines per cm or lines per inch – determines the resolution of the printing. Cell depth is generally measured in microns and, when related to the cell's opening, produces a depth-to-opening ratio. The most efficient producing a depth-to-opening ratio between 20 and 25 per cent.

Anilox tools are carefully selected for specific types of printing, substrates, and customer requirements. The printer may well perform test runs to determine the ideal anilox for producing the desired ink distribution for halftones, spot color and solids.

Chapter 8

Ancillary equipment for setting, measuring, testing, monitoring and adjusting tooling

Label and package printing, with all the associated finishing stages ranging through operations such as slitting, cutting, perforating, scoring, punching, foiling and embossing, involves some of the most sophisticated and precise technology found in the whole printing industry.

Tolerances are minute when setting-up equipment to precisely cut through a face material and adhesive but without marking or damaging a release liner. Similar precision set-up challenges arise when preparing for hot stamp foiling, or complex embossing using male and female dies. Where heat is also involved, as with hot foiling, settings may even change as dies and units heat up. Continuous monitoring and ongoing adjustments are then required.

Indeed, precision cutting, foiling, and embossing of labels and packaging can be considered as virtually impossible today without the use of pressure jacks (shown in Figure 8.1) or regulator gauges showing the amount of contact pressure being applied. Ideally, the aim should always be to apply the minimum cutting and pressure force so as to preserve tool life and bearings as much as possible.

Although experienced operators will undoubtedly have a good feel for how much torque/pressure is required for any specific tool to operate correctly and accurately, and exact measurement using pressure

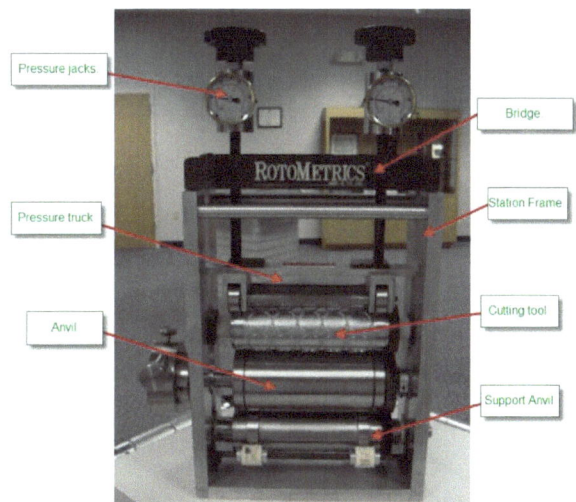

Figure 8.1 - A die-cutting or foiling unit with pressure measuring gauges/jacks

gauges will be the most consistent method for set up and longest tooling life possible. The printer can die-cut with optimum preload avoiding uneven cutting results. Documenting this pressure history can improve set up time for the next use of the tool as well as help to predict the life expectancy of the tooling.

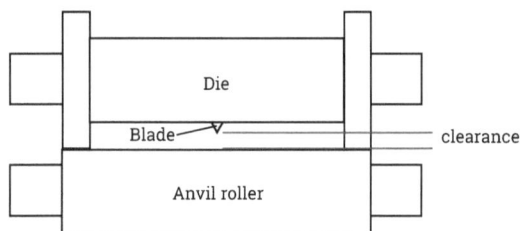

Figure 8.2 - Shows the clearance between the magnetic cylinder and the anvil roller which can be measured with a feeler gauge

FEELER GAUGES

For flexible die plates, the gap measurement between the base of the tool (surface at the magnets) is a critical specification for the manufacturing of the flexible die. This gap, also called 'air gap' or 'clearance' can be measured using a feeler gauge. This easy-to-use, precision tapered, measuring instrument is very accurate. Using a feeler gauge is quite easy; firstly, clean the magnetic cylinder and anvil roller carefully. Next, insert the magnetic cylinder into the cutting unit and apply standard pressure onto the magnetic cylinder. Finally, use the feeler gauge to measure the gap between anvil roller and magnetic cylinder to the left, in the center, and to the right to prove its parallelism. Then repeat the action after each third-revolution of the magnetic cylinder. These new measuring points not only indicate the real/actual slit dimension but also give information about the running accuracy and provide an accurate picture of the parallelism and gap between the magnetic cylinder and the anvil roller.

If the feeler gauge is used at regular intervals it will provide the operator with vital information about the condition of the die-cutting unit, pre-empting potential die-cutting problems. Using the information, a die can be made to suite the gap and type/thickness of the carrier.

However, when using rotary die-cutting systems, the basic requirement for perfect cutting is sufficient pre-tension between the magnetic/rotary cylinder and the anvil. Excessive or poor pre-tension may often lead to insufficient cutting results as well as unnecessary wear and tear with the printing machine. Different printing widths as well as different printing materials also require individual cutting pressures. In these cases more sophisticated pressure gauge systems, regulators and monitoring devices may be used. Some of the most commonly used gauges and devices are described below

PRESSURE GAUGES

Pressure adjusting and load cell monitoring equipment today – depending on the degree of sophistication – can enable press operators to set, monitor, adjust and detect or improve some or all of the following functions:

- Set a certain pressure from the start
- Ensure that the same and an even pressure is applied on both sides of the cutting unit (operating and drive side)
- Monitor the force required for rotary die-cutting or hot stamping
- Reduce the applied pressure to the minimum as soon as the machine reaches its optimum operational temperature
- Predict and prevent die failure
- Detect if cutting/foiling unit components (e.g. bearings) are failing
- Improve product quality and increase die life
- The compacting of load cells to allow easy installation

Some pressure gauge systems may also enable data to be downloaded, while other can include an alarm output with an adjustable setting point.

When assessing the technological aspects of pressure gauges, there are two key ways of monitoring the force or pressure being applied. These

two methods are:
- Conventional hydraulic pressure gauges
- Electronic pressure gauges

Essentially, both types of gauge do the same job.

HYDRAULIC PRESSURE GAUGES/JACKS

The simplest types of hydraulic pressure gauge systems enable machine operators to easily monitor and regulate the force being applied to cutting and other dies using easy to read gauges. They are therefore an ideal tool with which to help decrease downtime and extend the working life of cylinders, anvils and dies by reducing the amount of wear and damage caused by the application of excess pressure.

Pressure gauge systems may be already built into a new press, or they can be retrofitted to the majority of die-cutting, foiling or embossing stations at a later stage.

In use, the hydraulic indicator shows the internal pressure of the hydraulic fluid for a given load to the lead screw, thereby detecting the force being directly applied from the load to the die.

Figure 8.3 - RotoMetrics hydraulic pressure gauge system

In the example shown in Figure 8.3 the load cells and gauges are part of the screws, which eliminates the need for separate wiring or mechanical connection. They only require simple retrofitting of the locking device to the existing bridge in most cases. Screws can move freely up and down until the cam or clip lock is engaged, then pressure can be adjusted

and monitored. Easy to read clock dials or gauges, calibrated in pounds of force are certified to national standards.

Figure 8.4 - Universal pressure gauge. Source: Kocher+Beck

Another similar example of a hydraulic pressure gauge that replaces the exiting screw can be seen in Figure 8.4. This universal pressure gauge system is suitable for all rotary presses. The setting unit is mounted on a special, hardened threaded spindle, which is used in place of the original spindle the same as the one previously mentioned in Figure 8.3. It does not have to be laboriously screwed in and out as it is held in place by an ingenious rapid clamping system.

Figure 8.5 - Load cells (shown on truck). Source: RotomMetrics

Yet another style, shown in the load cells illustration in Figure 8.5, is universal and not specific to a press or unit. These are placed onto the pressure loading truck – they are a light-duty version. Also they may not be as easy to read as the gauges are not on top of the bridge as the others mentioned.

For all of these styles, pressure is applied straight to the cutting tool. Cutting pressure can be adjusted with the minimum of effort by means of a hand-wheel.

ELECTRONIC PRESSURE GAUGES
Electronic pressure gauges come in a variety of formats which offer active pressure control to maximize tool life and guarantee higher productivity.

In the active electronically controlled cutting pressure regulator show in Figure 8.6 a die-cutting head is mounted onto a stepper unit and is driven into position against an anvil cylinder by a servo motor to generate a specific pre-calculated cutting pressure.

Figure 8.6 - Schober electronically-controlled pressure cutting regulator

Feedback signals from the servo motor are constantly monitored to ensure that the cutting force between die cutting and anvil cylinder remains constant. This means that the cutting quality will not

be affected by fluctuation in the speed of the web. Due to the precisely controlled cutting pressure, tool wear can be reduced to a minimum and consequently lifetime of tooling is increased. In addition to the increased lifetime other advantages are: no scrap generated from bringing the machine down and back up, increased productivity, reproducible setting data and fast change-over.

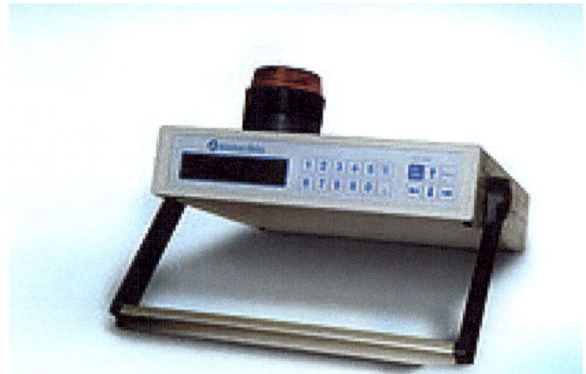

Figure 8.7 - Electronic pressure gauge

Another device for measuring rotary cutting pressure is that shown in Figure 8.7, which monitors the left and right side bearer rings separately to provide an exact measurement of actual cutting pressure, measured in kilo Newton (kN), and is

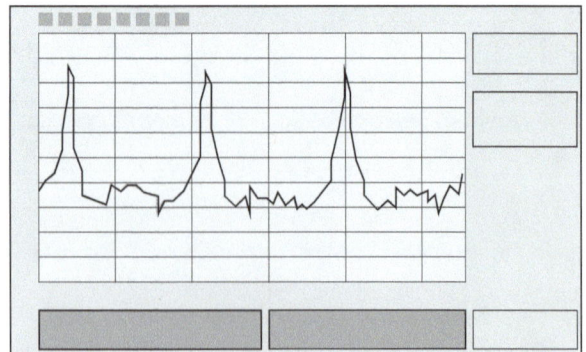

Figure 8.8 - Analysis of actual applied force (real time data)

designed to maximize tool life and increase cutting efficiency. The measuring gauge is equipped with an audible and visual alert feature to ensure the perfect cutting pressure. With its data interface option, this electronic power check can be linked into any operational data system and can offer superior cutting results and a means to avoid damage to magnetic cylinders and anvil rollers. It can be positioned on the pressure bridge or underneath the anvil roller.

As already mentioned, some electronic pressure gauge indicators will enable the analysis of real-time data showing the actual applied force at any given time in a revolution including the dynamic effects (see Figure 8.8.).

With all pressure gauges, regular calibration to National Standards is vital to ensure that instrumentation is operating at its peak performance and guaranteeing that the accuracy necessary is being delivered. Calibrating equipment periodically also has the added advantage of helping spot potential problems before they arise as instruments can lose accuracy for many reasons.

Figure 8.10 - This shows the RotoMetrics AccuStrike adjustable anvil roller

Firstly, to reduce the need to position the tool with fiber washers. Secondly, to eliminate the need to keep the die level as it is lowered into place. Thirdly — and perhaps most significantly — to allow the press operator to move a die in the across web direction with the turn of a knob. The blocks fit most presses without modification, and feature adjustment travel up to .250" / 6.35 mm. (Note: Washers are still needed to position the die gear from being adjusted over the anvil body.)

Figure 8.9 - Lateral adjustment journal blocks. Source: RotoMetrics

Figure 8.11 - A RotoMetrics adjustable anvil shown set up in the frame

LATERAL ADJUSTMENT JOURNAL BLOCKS

Lateral adjustment journal blocks (Figure 8.9) are used to replace standard journal blocks on the operator side of the press. They were designed with the press operator in mind by offering three key benefits.

ADJUSTABLE ANVIL ROLLERS

Instead of a fixed connection between the cutting cylinder and the anvil roller in which the bearers on the cutting cylinder are resting on the anvil under pressure, it is also possible to have an 'adjustable' anvil roller. See Figures 8.10, 8.11 and 8.12. This approach changes the effective undercut in order to compensate for variations or deliberate changes in thickness/calliper of the liner and to some degree, the wear of the cutting edges.

Figure 8.12 - Shows a Kocher+Beck adjustable anvil roller installed in a cutting unit.

In these systems, the bearers of the cutting cylinder are resting on separate bearers on the anvil. The bearers on the anvil are no longer connected to the anvil body, but the anvil body is instead spinning freely on a cam.

When the cam is turned, the gap between the cutting cylinder and the anvil body can therefore be adjusted in small increments to suit either the wear on the cutting tool or variations in material.

Some of the anvil cylinders that are adjustable are no longer solid so risk of deflection increases slightly. This is typically compensated by an increase in diameter. It is also crucial that the operator remembers to reset the gap when a new cutting tool is placed in the cutting unit or it might be damaged straight away. The adjustments can be made during operation and do not require the cutting process to stop. Some of the designs available are easy 'drop-in'

ready and require little or no maintenance.

There are also less sophisticated adjustment possibilities. One possibility is a 'compressible' bearer. These bearers are made with a deflection zone that allows the pressure applied by the lead screws to 'compress' the bearers. The compression range is however limited compared to the mechanical systems.

It is also possible to choose a more traditional method of exchanging the 'zero' anvil cylinder with a 'plus' or 'minus' anvil cylinder. A 'plus' anvil cylinder is an anvil cylinder with a bigger diameter in the cutting zone than that in which the bearers of the cutting cylinder are resting, thereby closing the gap between the cutting surface and the blades on the cutting die.

Alternatively, a 'minus' anvil cylinder will have a smaller diameter than the bearers in the cutting zone of the anvil cylinder, putting the cutting blades farther away from the material (reducing the cutting depth). Using an anvil with a step in the body (a plus or a minus step) will require the cutting unit be at a standstill during the change out.

Figure 8.13 - Examples of correction tools. Source: Electro Optic

CORRECTION TOOLS

Special tools have been developed for label converters to be able to adjust the height of the cutting edge and to repair minor damages to the cutting lines/blades during or after the use of flexible magnetic dies. A plus tool is used to increase/lift the height of the cutting edge, while the minus tool enables the cutting line to be re-sharpened to the original cutting angle.

Chapter 9

Inspecting, cleaning, handling, storage and safety considerations

Any operations that involve the handling, setting up, cleaning, inspection or storage of cutting, embossing, foiling, sheeting or perforating tools – even cylinders, anvils and support rollers – has the potential to either damage the tooling or cause a safety or health risk to the workforce. This applies whether the tooling is flatbed, solid rotary or flexible. Some operations and tools however, are perhaps more likely to cause damage or safety risks than others.

In particular, the manufacture and use of tooling at the finishing end of a roll-label press often involves engineered products that can be heavier, harder to handle, bulkier and potentially more likely to be damaged or cause damage than in many other label production applications. Quite simply, moving and handling precision engineered tools during manufacturing or in warehouse, production or storage areas requires specific expertise and training.

This chapter therefore aims to look at some of the main areas in pressure-sensitive label manufacturing – and in particular at the finishing end of the press – in which safety or damage risks have the potential to most likely occur. These areas and operations can include the following:

- Solid rotary cutting, embossing or foiling tools, as well as cylinders and anvils, can be quite heavy and a challenge to lift and move in or out of the converting line and handle safely without risk to the tool or operator
- Flatbed, rotary or flexible cutting dies have very sharp edges. Operators can sustain cuts and the cutting edges can be damaged during handling, set-up or adjustment
- Unpacking of incoming cutting dies and other tooling – and re-packing for storage or shipping – again has the potential to cause damage to the tool or operator during handling
- Cleaning and treating (oiling) prior to storage may also lead to handling or tool damage if not carried out carefully
- Operators may adopt awkward postures of back, neck and arms when inserting tooling or making machine adjustments, cleaning, and performing other tasks on the finishing line
- Loose object or hand tools, Allen keys, etc., left lying around the press may fall or be dislodged into machine working areas and cause serious tool or machine damage
- Items of clothing (hard metallic buttons) or jewellery (rings, watches, chains) worn by the operator during handling, installation or set-up may cause scratches, nicks or abrasions on precision tools, and possible injury to the operator

- Insufficient attention to set-up and running tolerances and pressures may lead to unit or tooling damage
- Poor tooling maintenance and storage conditions can lead to deterioration in tools over time.

There are always other areas where improved care and attention to tooling may be required, depending on the particular manufacturing or factory circumstances, but regular recording of the circumstances of tooling damage or operator safety issues should enable a converter to make further recommendations for handling, storage, or operator training.

The handling of tooling should not be just a secondary concern. The systems and procedures that are used to move and handle tooling as it enters, passes through, and departs manufacturing processes, machinery or fabrication areas can be a critical key to company productivity. The personnel that handle tooling need specific handling and usage instructions.

Often this training may be overlooked or considered secondary to the process of the machine set up. Good storage, transportation and handling systems used in label converting operations can significantly reduce costs, increase productivity and create a safer, more ergonomic production environment.

Some of the key factors that need to be considered and addressed in the manufacturing and subsequent use of tooling in the label production plant are:

- Ensuring that there is no transportation damage, either from or to the manufacturer or within the converting facility
- The elimination as far as possible of any form of tool damage
- The reduction of production time and costs through the use of optimum handling, storage and usage procedures
- The provision of easy access to each individual tool
- The elimination or minimizing of any heavy or awkward lifting

- Obtaining a significant decrease in the chance of accidents

Quite simply, the aim should be to make life easier and safer for employees throughout the whole production environment and to minimize the risk of damage to precision engineered tooling and production machinery. A key step to achieving these aims is to implement improved training procedures and to better educate workers in how and where tooling damage is most likely occur.

TRAINING
All press operators and other employees involved in the handling and use of any form of tooling should be properly trained and competent. All training given should be based on relevant safe systems of work and should include a good working knowledge of the specific tooling to be used, it's installation and set-up, how it should be handled safely, the operation of the press and converting units, the function and correct use of all the necessary controls, how to complete daily checks and what to do if any part of the machinery or tooling fails or becomes damaged. There should also be guidance on what the employee or operator needs to do if there are any concerns.

Training should also clearly identify which tasks an employee or operator should not carry out (i.e certain types of press or tooling repairs) as part of a wider set of 'dos and don'ts' incorporated into their training program. Retraining or refresher training for existing staff is also important if new safeguarding methods are introduced, and is particularly important if the safe systems of work are changed while the machinery stay the same.

Managers and supervisors have a crucial role in ensuring that safe handling, set-up, adjustment, maintenance and storage procedures are followed and that all operators and other employees follow agreed safe systems of work. Managers and supervisors also need to be competent to carry out their safety duties effectively and may themselves require further or refresher training where there are any changes to working procedures, systems or equipment.

GENERAL SAFETY AND HEALTH GUIDELINES

General principles for the avoidance of accidents in and around label and converting units and presses include:

- Know all the control points around the machine and how to stop in an emergency
- Check all guards are in position and working properly before starting the machine
- Wear suitable clothing. No loose clothing, ties, cuffs, etc., which may get caught up. Wear good boots or shoes
- Watch for oil, grease, ink or varnish around the machine and clean it up
- Keep workbenches and floor areas tidy and free from waste
- Check that tools and equipment are not lying on the machine when starting up
- Stop the machine before making adjustments and always use the correct tools for the job
- Never try to move obstructions or blockages while the machine is still in motion
- When working on a machine make sure that the power is shut off in such a way that it cannot be started accidently
- Keep all gangways and passages around machines clear of obstruction. Even small items may cause a fall
- Ensure all waste substrate is placed in the proper containers.

HANDLING GUIDELINES

Many printing industry accidents occur each year due to poor lifting, moving, handling, installing or adjusting of everything from reels, matrix waste, pallets, ink containers or machinery equipment. Yet the basics of lifting and moving heavy or awkward shapes are not always well explained. This point needs to be emphasised when lifting and handling cylinders, anvils, solid rotary cutting, embossing and foiling dies and tools, even flat dies, to ensure that back, arm, shoulder and fingers are not strained or damaged during lifting and moving. In addition, it is essential that cutting blades are protected at all times while handling, loading/unloading die-cutting, sheeting or perforating tools.

Fundamental points to remember in the manual handling and lifting of heavy dies or awkward or difficult shapes are as follows:

- Never persist in attempting to lift loads or heavy tools which cause a feeling of strain
- Get a good grip with the palm of the hands when lifting and handling, not just fingers only. Use protective gloves where thought necessary
- Do not stand holding heavy tools or loads. Rest on a bench or platform of a suitable height if any delay is unavoidable
- Never change the grip while carrying
- Use handling aids or hoists wherever possible
- Train workers to keep lifts below shoulders and above knee height
- Educate workers about risks to the low back related to handling and twisting.

It should be noted that the design of rotary tooling and converting units may typically require an operator to bend at the waste, twist and reach out to position, set and adjust cylinders, rollers, guides, rotary knives, etc. Workers of course, do not usually hold these postures for any significant duration in their non-working life. The frequency that workers on a converting line adopt these postures will depend on the particular product mix being processed and the make, model and year of the rotary die-cutting, embossing or foiling unit.

Factors that may increase the risk of injury when carrying out this type of operation include:

- Duration of time in which awkward postures may ned to be held
- The frequency of adopting awkward postures
- The number of extreme postures adopted
- The accessibility to adjustment areas (machine design).

AUTOMATING THE HANDLING PROCESS

Some narrow-web press manufacturers have now incorporated innovative quick-change die-cutting units for very fast job change-overs. They are designed to save time on each job-setup, with easy handling of tools and less material waste.

These units feature an operator-friendly make-ready position, enabling the operator to set-up a magnetic base cylinder and flexible die while the press is running. When the press is stopped, the previous cylinder is rolled to an unload position and the new cylinder is rolled into the running position – a process that takes no more than a few seconds.

The magnetic cylinders are equipped with bearing blocks that are automatically engaged and retained in the QC-station. Combined with precise tooling and automatic hydraulic loading, this allows the operator to change dies in seconds.

Other manufacturers have developed solutions such as a unique cassette and cart combination which allows the operator to load the die off-line and then conveniently load it into the die station without the use of a hoist. Dies can be changed in a little as 30 seconds. Also available are special mobile roll/ cylinder lifts that have been designed for lifting, handling and rotating, everything from self-adhesive laminate roll stock to finished printed reels and dies. Such lifts can be equipped with multiple attachments, so providing a turn-key solution for a variety of converter handling needs.

PRE-PRODUCTION PROCEDURES

Prior to tooling being used it is advisable to go through procedures prior to loading the tool or dies in the press so minimize the potential for problems. In particular, all dies and cylinders – whether for cutting, embossing, foiling or sheeting/perforation – should be carefully checked and inspected for any visible signs of damage.

Where appropriate, all cylinders, rotary dies, and tools should be examined to ensure that gears are attached properly and secure. Setting-up of rotary dies will normally involve hand peeling a full repeat, pulling waste matrix and carrying out an ink stain test (as explained in Chapter 3).

At all times dies and other tooling needs to be handled carefully, especially during loading, unloading and cleaning.

SETTING UP

After installing cutting, embossing or foiling dies it should be routine to carry out a check to ensure that they have been installed correctly, as well as undertaking all the necessary approval or quality

checks before full production commences.

Press tolerances should be set accurately and checked periodically. When adjusted accordingly, it is possible to considerably increase the life of cutting, foil stamping, embossing and debossing dies. Also, be careful in make-ready. Scratching and dents can be avoided with just a little attention. With proper handling in all of these areas, it is possible to get the maximum life and usage from almost any type of die.

Figure 9.1 - Use manufacturer's shipping boxes for safe storage. Source: RotoMetrics

STORAGE OF CUTTING, EMBOSSING OR FOILING TOOLS

Unless they are only used for a one-off or specific operation, most types of tooling will require effective cleaning and proper storage after use so that they can be used again at a time in the future. Steps to take should include:

- Inspecting the tool for any visual signs of damage that may need the die to be returned to the manufacture for remedial treatment before storing
- Clean or gently brush off any dust, hairs or other debris from the surface of the particular tool, ensuring that it is handled safely. Avoid any possibility of damage to the surface of cutting, embossing or foiling edges or faces
- Protect the tool from moisture by spaying with a light coating of oil
- Use chemically treated wrapping tools
- Wrap the tool in a bubble pack
- Use the manufacturers shipping boxes for safe storage when possible

- Support rotary dies in blocks
- Ensure that nothing in the box comes into contact with cutting blades or the surface of embossing or foiling tools.

Brass, copper, or magnesium foiling or embossing flat dies can easily be scratched or nicked, so it is always a good idea to store flat dies in a foam envelope or bubble wrap of some sort. In addition, it is always recommended to store flat dies standing on end (vertical) rather than stacking them on top of one another. Some type of filing system, such as a filing cabinet with drawers, is best as opposed to stacking them up in a box or on a shelf.

For embossing/debossing flat dies that also have a pre-cast counter, it is recommended to re-pin the counters to the dies before storing them. Do not simply tape the counter to the die, since the die and counter can move or shift and cause wear on both the die and the counter. These wear or abrasion marks certainly will appear on press if the dies are not properly stored.

It also is recommended to consider handling copper and magnesium flat stamping dies differently than more expensive brass embossing dies, purging flat stamp dies in, say, two years and keeping brass dies for three to four years. If a flat stamp job is requested again after two years, it is much less costly to replace the flat stamp dies if the customer insists that they should not have to pay the cost to duplicate them.

Where flat dies are stored in large envelopes and filed vertically, then the layout or cut sheet from the job, plus the original Purchase Order information and a proof of the final product can be kept in this envelope, thereby storing the dies and keeping all of the job information in one place. If film was used with the job, filing this also is a good idea. Many times, the film is useful when registering the die to print or foil if the job is repeated.

In addition, it is always a good idea to make sure a die number from the manufacturer is kept in the storage envelope, storage box or file if it is not included on the die. It is much easier for the die manufacturer to make new dies, new counters, or even counter masters from old combination dies if it can reference this number.

DISPOSAL OF OLD DIES AND TOOLS

The recommended storage time for dies is usually between two and four years, depending on the tooling material used and the manufacturer's recommendations.

Copper, brass, and magnesium dies are recyclable and most communities have a local metal recycler. Usually the price for recycled brass and copper is at a premium and the payback on these metals can be quite high. This should provide even more of an incentive to purge current die inventory and create an organized storage system.

CONCLUSIONS

Knowing the best procedures and practices to handle cutting, foil stamping and/or embossing dies has always been something of a gray area in the industry. It is hoped that this chapter has explained and clarified key aspects of the handling, storage, inspection and cleaning of label industry tooling, as well as creating some standard guidelines that can help manufacturers and converters to become more organized and communicate more efficiently with their customers.

Chapter 10

A guide to troubleshooting when using label dies and related tooling

Throughout this book it has been emphasized that the use of cutting, foiling or embossing dies – as well as all the related cylinders, anvils, pressure units, slitters and gauges – is a complex process using a wide range of precision engineered tooling. Careful pressure settings and adjustments are important for success as well as correct choice of cutting angles and cutting profiles, and attention to all aspects of handling and storage.

Get all of this correct and it is hoped that the highest quality performance and optimal results will be consistently achieved. Yet problems can still occur. In general most problems can be divided into one or more of the following categories. These categories are shown in Figure 10.1.

Within each of these three categories there are then a whole variety of parameters to consider when looking to troubleshoot and solve problems. Parameters as diverse as a damaged die, polluted or worn bearers, variations in liner thickness, incorrect web tension, excessive pressure on the pressure

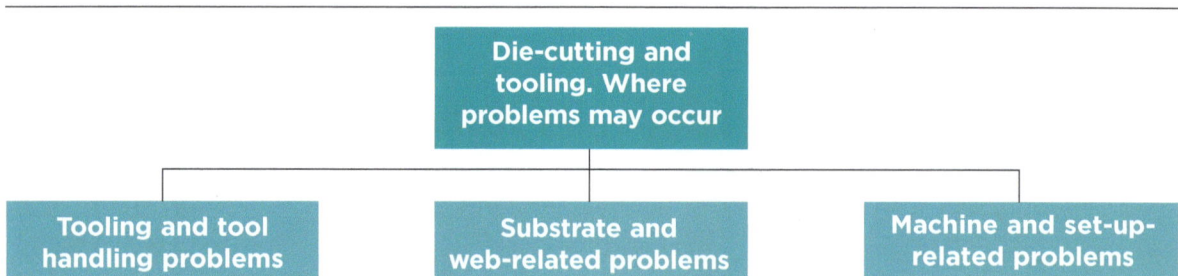

```
           ┌──────────────────────────┐
           │   Die-cutting and        │
           │   tooling. Where         │
           │   problems may occur     │
           └──────────────────────────┘
      ┌───────────────┼───────────────┐
┌──────────────┐ ┌──────────────┐ ┌──────────────┐
│ Tooling and  │ │ Substrate and│ │ Machine and  │
│ tool handling│ │ web-related  │ │ set-up-      │
│ problems     │ │ problems     │ │ related      │
│              │ │              │ │ problems     │
└──────────────┘ └──────────────┘ └──────────────┘
```

Figure 10.1 - Categories of die-cutting and tooling problems

bridge, deflection of the rotary die or magnetic cylinder, stripping angle too sharp, or incorrect layout.

As can be seen, many factors may be involved. It is therefore necessary to try to work out what is happening. Use a process of deduction or analysis; work out what you can see from studying the tool, bearings, cylinder, web or pressure gauge. Assess which of the three categories in the chart may be the most relevant or applicable to the specific problem that is occurring. Narrow things down as much as possible and, above all, do not attempt to guess.

To aid the troubleshooting process – both for the operator and for the tooling supplier – proper die-cutting records should be kept. These should identify:

- The date(s) the die has been run
- The number of revolutions run on each date
- The date(s) the die has been returned for repair
- A description of the material being cut (type, caliper, liner) during each run
- The starting and ending pressure during each run.

For the press operator, there are then a number of first steps that can be taken if the die is not cutting correctly, including:

- Making sure the die was made for the material being cut. Check the liner thickness to make sure that it is uniform. If the die has been made for a different material, then a stepped anvil may help
- Checking the edge of the label for hairs. If it is cut cleanly, the problem may be the adhesive. Again, check that the correct material is being used. Also check to see whether the die is bouncing or flexing. Adjust the pressure if necessary
- Check to see whether the correct magnetic cylinder is being used. Change if necessary
- Establishing whether the die stopped cutting from one instant to the next or gradually over time. If so, the anvil roller or magnetic cylinder may be worn. Try a different anvil. Proper lubrication will also greatly reduce the wear of the bearers of the die and the anvil roll surface.

Indeed, lubricating and wiping of the bearers will reduce the friction that causes thermal expansion. Thermal expansion of the bearers will cause the cutting blades to lift away from the anvil roller and can lead to fuzzy cutting or causing the labels to lift with the waste matrix.

Having set-out some basic troubleshooting pointers, let's now look in more detail at the key problems that may occur during die-cutting process and assess the nature and possible troubleshooting solutions.

WASTE MATRIX BREAKING (REPETITIVE)

Damaged rotary die / flexible die. Check the cutting lines for possible damage. If a rotary die is damaged, send the die back for repair. In case of damaged cutting lines on a flexible die, use a spare flexible die or order a new one. Repairing of small areas of damage is possible by making use of special repair tools.

Worn cutting cylinder / flexible die. Try to finish the job with a plus (corrected) anvil roller or use an adjustable anvil roller system. For flexible dies, you can use a thin film (10-15 μ) on the backside of the flexible die or a light coating of spray paint all over the backside or just in the area not cutting. This will lift the die slightly and may enable the job to be finished. In all cases, do not forget to reduce the pressure on the pressure bridge. Flex dies cannot be re-sharpened but some small areas of damage may be repairable with a special tool. Solid rotary cutting dies can be re-sharpened and small areas of damage may be repaired. After the job is finished, send in the rotary die for repair or to be replaced.

Polluted bearers. Check the die for pollution. There may be a build-up of ink, adhesive, paper or film particles that are interfering between surfaces in contact with each other.

Worn bearings. The condition of the bearings should be an integral part of the standard machine maintenance program. Check the bearings in the bearing blocks and replace if necessary. Check for excessive wear or lack of lubrication.

Incorrect position of the label shape in relation to the web direction. Check the position of the label shape and if possible adjust this to prevent the waste from breaking. See Figures 3.10 and 3.11

in Chapter 3 for more information.

Waste matrix too narrow. Check the width of the waste matrix (gap across and around) on all positions as well as the waste along the outside edges.

DAMAGING OF LINER (RANDOM)

Pollution of anvil roller. Check the anvil roller for adhesive build up and paper/film particles.

Damaged anvil roller. Check the anvil roller for damages. If the anvil roller is damaged, send the anvil roller back to the tooling supplier for repair or order a new one.

Irregular thickness of the liner. Measure the caliper (thickness) of the liner and compare this with the specifications of the material supplier. These measurements are indicative, as the caliper of paper is influenced by temperature/relative humidity.

DAMAGING OF LINER (REPETITIVE)

Pollution on the back of the flexible die. Clean the back of the flexible die carefully before mounting the plate to the cylinder.

Pollution / damage on the magnetic cylinder. Clean the magnetic cylinder carefully before putting on a flexible die. If the magnetic cylinder has been damaged or is malfunctioning (e.g. lifting or sinking of magnets), send the cylinder back to the manufacturer for repair. If repair of the magnetic cylinder is not possible or not economical, order a new magnetic cylinder from the tooling supplier.

Worn bearers; magnetic cylinder / rotary die. Send the rotary die or magnetic cylinder in for repair or order a new cylinder.

LABELS GO UP WITH THE WASTE MATRIX (RANDOM)

Worn cutting cylinder / flexible die. Try to finish the job with a plus (corrected) anvil roller or use an adjustable anvil roller system. For flexible dies, you can use a thin film (10-15 μ) on the backside of the flexible die or a light coating of spray paint all over the backside or just in the area not cutting. This will lift the die slightly and may enable the job to be finished. In all cases, do not forget to reduce the pressure on the pressure bridge. Flex dies cannot be

re-sharpened but some small areas of damage may be repairable with a special tool. Solid rotary cutting dies can be re-sharpened and small areas of damage may be repaired. After the job is finished, send in the rotary die for repair or to be replaced.

Liner not correctly specified. Try to finish the job with a stepped anvil roller or use an adjustable anvil roller system to compensate for the error. If the liner is too thin, try to use a thin film (10-15 μ) on the back of the flexible die. Finally, order a new flexible die with the correct specification or send in the rotary die to determine if the necessary adjustments can be made to the existing die.

Variation in liner thickness (calliper). The thickness (calliper) of a liner can be measured by making use of a micrometer. If the variation in liner thickness is too large, send the material to the material supplier for further inspection and evaluation.

Incorrect lay-out for stripping the waste. If possible, change the lay-out or call the tooling supplier for advice.

Material between bearer and anvil roller. Check the web width and its position between the bearers.

Damaged, polluted or worn anvil roller. Check the anvil roller. If the anvil roller is worn and/or damaged, use a back-up anvil roller. If a back-up anvil roller is not available, send in the cylinder for repair or order a new anvil roller. If the anvil roller is polluted, clean the cylinder.

LABELS GO UP WITH THE WASTE MATRIX (REPETITIVE)

Damaged/worn cutting lines. Check the cutting lines for possible damage. If a rotary die is damaged or worn out, send the die back for repair. In case of damaged cutting lines of a flexible die, use a spare flexible die or order a new one. Repair of a small area of damages on a flexible die may be possible by making use of special repair tools.

Pollution of cutting cylinder / flexible die. Check the die for pollution, adhesive build up and paper/film particles.

Polluted bearers. Check the die for pollution, adhesive build up and paper/film particles.

WASTE MATRIX BREAKING (RANDOM)

Pollution of anvil roller. Check the anvil roller for pollution, adhesive build up and paper/film particles.

Liner not correctly specified. Try to finish the job with a corrected anvil roller or use an adjustable anvil roller system to compensate for the error. If the liner is too thin, try to use a thin film (10-15 µ) on the back of the flexible die or very light coating of spray paint. Finally, order a new flexible die with the right specification or send in the rotary die for making the necessary adjustments.

Variation in liner thickness (calliper). Measure the thickness (calliper) of the liner with a micrometer. If the variation in liner thickness is too large, send in the material to the material supplier for further inspection and evaluation.

Worn bearings. The condition of the bearings should be an integral part of the standard machine maintenance program. Check the bearings in the bearing blocks and replace if necessary. Check for excessive wear or lack of lubrication.

Incorrect web tension. Check the web tension and if possible adjust.

Stripping roll too small or stripping angle too sharp. Check diameter of stripping roll and adjust this if possible. If necessary, change the stripping angle under which the waste matrix is being stripped.

DAMAGING OF LINER (REPETITIVE)

Damaged rotary die / flexible die. Check the cutting lines for possible damages. If a rotary die is damaged, send the die back for repair. In case of damaged cutting lines on a flexible die, use a spare flexible die or order a new one. Repairing of small areas of damage is possible by making use of special repair tools.

Stepped anvil roller. Check if by mistake a stepped (plus) anvil roller is being used or if an adjustable anvil roller needs to be reset to a zero position.

Incorrect calliper of liner. If the liner is too thick, the cutting lines will cut too deep into the liner, thus damaging the silicone layer. Use a stepped (minus) anvil roller or use an adjustable anvil roller system to compensate for the difference. Finally, order a new flexible die with the corrected specification or

send in the rotary die for the necessary adjustments to be made.

Too much pressure on the pressure bridge. Reduce the pressure on the pressure bridge and reset, then start again.

HIGH PRESSURE ON THE PRESSURE BRIDGE

Worn cutting cylinder / flexible die. Try to finish the job with a plus (corrected) anvil roller or use an adjustable anvil roller system. For flexible dies, use a thin film (10-15 µ) on the backside of the flexible die or a light coating of spray paint all over the backside or just in the area not cutting. This will lift the die slightly and may enable the job to be finished. In all cases, do not forget to reduce the pressure on the pressure bridge. Flex dies cannot be re-sharpened but some small areas of damage may be repairable with a special tool. Solid rotary cutting dies can be re-sharpened and small areas of damage may be repaired. After the job is finished, send in the rotary die for repair or to be replaced.

Flexible die too low / drop too small (rotary die). Flexible die height or the drop (distance between bearer and top of the cutting line) of the rotary die do not match the liner thickness.

Incorrect liner thickness (caliper). Check if the liner thickness is correct.

Stepped anvil roller. Check if by mistake a stepped (minus) anvil roller is being used or that an adjustable anvil roller needs to be reset to a zero position.

Front material too thick. Check that the thickness of the front material (compressed) does not exceed the actual height of the cutting line.

Pollution of the bearers. Check the die for pollution, adhesive build up and paper/film particles.

NO (BAD) CUTTING IN THE MIDDLE OF THE WEB

Deflection of the rotary die / magnetic cylinder. If the diameter of the rotary die / magnetic cylinder is too small this may lead to bending (deflection) of a cylinder (see Figure 3.6 in Chapter 3). If possible, change the layout of the flexible die or use cylinders with a larger repeat size.

Deflection of the anvil roller. If possible use an

anvil roller with a larger repeat size.

CUTTING TOO DEEP IN THE MIDDLE OF THE WEB

Aligning, centreline. Check if the rotary die/ magnetic cylinder and the anvil roller have been aligned correctly.

Worn bearings. Check the bearings in the bearing blocks and replace them if necessary. (Checking of bearings should be integrated into the maintenance program of the machine).

Shaft diameter. Check if the shaft diameter corresponds to the inner diameter of the bearings in the bearing blocks.

CUTTING ONLY ONE SIDE

Wrong gear. Check the pitch of the gear if the magnetic cylinder or rotary die is not cutting on the gear side.

Damaged/worn bearers. Check if the bearers of a rotary die or magnetic cylinder are worn or damaged. To check the cutting result, try to put the flexible die onto another magnetic cylinder. If the bearers are worn or damaged, send in the cylinder for repair or order a new one.

Pollution of bearers. Check the die for pollution, adhesive build up and paper/film particles.

INCORRECT LABEL DIMENSIONS

Wrong specifications on tool. Compare the specifications of the label (order form) against the order acknowledgement or packing slip. (The solid rotary or flexible tool will arrive with samples in the box but will demonstrate the cutting performance only. They are not accurate for size.) If the specifications/dimensions of the flexible or rotary die are incorrect, contact the tooling supplier for resolution.

Incorrect web tension. Tension in the web can affect the size of the label in the web direction. If the label size in the web direction is too small, tension may be too high. If the label size is too long, then the tension may be too low.

FINALLY...

If it still not possible to solve the particular die-cutting or tooling problem, then it may be time to get in touch with the tooling supplier.

Chapter 11

Glossary of die-cutting and tooling terminology

The aim of this book has been to provide a good basic understanding of the use and production of the many different forms of die-cutting and tooling that a label converting plant will come across in their day-to-day business. However, there are many other words and terms used by various industry suppliers that converters may come across when reading technical articles or supplier literature.

It has therefore been decided to include a supplementary 'Glossary of die-cutting and tooling terminology' in the book so that readers and students can have a quick reference source to go to when they have a query. The Glossary is not meant to be fully comprehensive, but combined with the various book chapters, should nevertheless cover the great majority of words and basic terms that they may come across on a regular or periodic basis.

Suppliers are welcome to submit further terms so that these can be added to the listings below in subsequent publications.

AN ALPHABETICAL GLOSSARY

Abrasiveness – The tendency of some papers, surface coatings, treatments or inks to abrade or wear away die edges, slitter wheels, printing surfaces, etc., by friction.

Air eject cylinder – Developed as an alternative to the more expensive male and female cutting systems for long runs and the solid rotary die-cutter with rubber inserts and without any waste control. The system is making use of so-called air forks, to blow air (6 – 8 bar) into air channels. These air channels, are linked to bores in the cavities. Die-

cutting of the holes and blowing out of the waste into a stainless steel vacuum box is done simultaneously.

Anvil position – The cutting cylinders (rotary die, flexible die, etc.) is placed in the bottom position of the cutting unit, whereas the anvil roller is placed in the top position of the cutting unit.

Anvil roller – Hardened steel roller upon which the bearers of a rotary die, magnetic cylinder, perforation cylinder, etc., run. Normally, this cylinder is placed in the bottom position of a die station. However, for certain jobs it is necessary to place the anvil roller in top and the cutting cylinder in bottom position. In case of a support roller the anvil roller would be in the middle position.

Anti-stick coatings – Coatings, engineered to reduce the ability of other materials to stick to them. For certain applications such as hot-melt adhesives, multi-layered labels and labels with high application of adhesive, tooling suppliers will offer the possibility to coat the dies. These coatings should prevent adhesive and ink adhering to the cutting edge and adhesive bridging taking place between the label and the waste matrix.

Axial – Consider a cylinder. The top of the cylinder is a circle. A radius is a line drawn from the centre of the circle to the border or circumference of the circle. Movement parallel to a radius is called 'radial movement'. The axle of a metal cylinder with shafts on each end may have movement parallel to the axle, that is, from one shaft to the other. This is called 'axial movement'.

Axis – Axis of revolution of the gear; center line of the shaft.

Backlash – Is the striking back of connected wheels in a piece of mechanism when pressure is applied. Another source defines it as the maximum distance through which one part of something can be moved without moving a connected part. In the context of gears backlash, sometimes called lash or play, is clearance between mating components, or the amount of lost motion due to clearance or slackness when movement is reversed and contact is re-established. For example, in a pair of gears backlash is the amount of clearance between mated gear teeth.

Bearings – The hard rings, often including balls or rollers, which provide a smooth rotary movement to cutters and printing cylinders that come into contact with each other during printing and die-cutting.

Bearing block – A block with internal bearings that holds the cylinders (magnetic cylinder, rotary die, anvil roller, etc.) in position in a die-cutting station.

Butt cut labels – Rectangular labels in a continuous web, which are separated by a single knife cut through the label and/or liner across the web. No matrix is removed between the labels, as is the case with die-cut labels. Separated individual labels on a backing liner are applied by hand or to its application to a container, product or surface.

Back slit – A cut in the release liner or backing, usually along the web, but can be on the back of sheeted pressure-sensitive laminate to allow the face stock to be easily peeled away by hand when die-cutting has not be used.

Case hardening – Case hardening or surface hardening is the process of hardening the surface of a metal object while allowing the metal underneath to remain soft, thus forming a thin layer of harder metal (called 'the case') at the surface.

Cavity – Any enclosed shapes on a rotary/flexible die.

Cold-foiling – A more recent development of hot-foil blocking, the cold foil process makes use of a print unit to print a special adhesive on the label web where a metallic effect is required. When the metallic foil is brought into contact with the adhesive it adheres to it to produce the printed foil design on the label.

Converting process – in package or label production, converting covers any process performed to manufacture a complete package or label from a raw material or an unfinished material. The process of transforming rolls of self-adhesive material into labels on a release liner (carrier) presented in rolls or sheets, so as to enable end users to apply them to products, packages or surfaces. This process includes slitting, die-cutting and matrix stripping operations, sheeting, and may also include printing.

Cross cutting – Cutting across the web.

Cross perforations – Perforations across the width of a continuous web for easy separation of individual sheets and/or fan folding of continuous labels. Also any perforation applied at right angles to a label or page, depending on its printed format.

Crush cut – A cut made using a rotary blade in contact with an anvil or base roll.

Crush cutting – Commonly used for adhesives and some papers. The score is a circular blade cutting against a hardened steel anvil. The circular knife then ´crushes´ the web against the anvil. Generally, slit edges are not of the highest quality, but crush cutting makes up for it in quick and easy set-up procedures.

Cutter bevels – In die-cutting, the angle of the material supporting the peak of the cutter. The smoothness of the bevel sides directly affects the amount of pressure that is required to penetrate the laminate surface – see cutting angle.

Cut & Tie – The term used when describing a perforation. The cut penetrates through whilst the tie remains to hold the material together.

Cutting station/unit – The part of a label press that contains the equipment to cut a shape or pattern into a given material. A cutting unit includes an anvil roller and/or support roller rotatable on a machine frame about an axis of rotation. The anvil roller has an anvil surface. A cutting tool is mounted on the machine frame for rotation about an axis of rotation, with a cutter interacting with the anvil surface and with supporting rings which are held on the cutting tool and support it relative to the anvil roller with their supporting ring surfaces and/or vice versa.

Cut-to-shape – Is originally a philatelic term referring to a postage stamp or postal stationery (printed stamp image) that has been cut to the shape of the design, such as an octagon, circle or oval, instead of simply cut into a square or rectangular shape.

Die-cutting – The process of cutting a label shape with a die. Most self-adhesive labels and some wet-glue and in-mould labels have to be die-cut to shape as part of their manufacturing and finishing procedure. Depending on the type of label and the printing and/or die-cutting requirement, the operation may be performed using high or hollow dies, flat dies, rotary dies, flexible dies or most recently with digital die-cutting (laser cutting).

Dial gauge – A dial gauge is a precision measurement commonly used to measure machined parts for production tolerances (run out). Dial gauges are capable of producing extremely fine measurement values. Plunger instruments are generally used in conjunction with a clamp or stand which holds the gauge in a fixed position in relation to the work piece. The work piece is then rotated or moved to take the measurements.

Die life – Meterage expected from a new die or that expected following a re-sharpening of a die. Estimates of die life depend on machine setting, type of label-stock, type of adhesive and on operator handling. Estimates of meterage from a die may vary considerably from company to company, machine to machine or job to job.

Die line – A blueprint. Drawing or computer-generated layout of the cutting shape or shapes of a die. Maybe supplied with artwork as an overlay, as a blue line on the base artwork or supplied as computer-generated data on disk, CD or transmitted electronically.

Die strike – The impression left on a backing liner after being converted by a cutting die.

Die wipe – A test to check the evenness of a rotary or flexible die cutting pattern on a backing liner.

A solvent pen/marker or fine colored powder is wiped over the silicone surface where a die impression is evident. Where the die has penetrated the silicone coating, the ink or powder stains the backing highlighting unevenness in the cutting depth which can cause the waste stripping matrix to break (see FINAT test method 23 + 23b).

Distortion – A change in the dimensions of an object to compensate for change in length when a flexible die is wrapped around a magnetic cylinder (to be calculated by your tooling supplier.

Dual height – Two different cutting heights combined in 1 tool, possible in both solid cutting tools and flexible dies. In a die drawing, it should be clearly indicated which cutting lines are cutting the different layers of material.

EDM dies – Rotary cutting dies produced using electronic discharge machining (EDM) by eroding the cutting lines. The hardening process takes place before the EDM process thereby eliminating possible distortion of the cutting lines/image. Consequently, these type of rotary dies generally will have a higher accuracy than milled dies.

Ejection rubber – A variety of materials used for facilitating the flatbed and rotary cutting process. These materials used vary in thickness, structure and hardness (shore). It is important to use ejection rubber that is slightly higher than the actual height of the cutting line. This will cause compression (sufficient energy and ejection force) on the rubber thus ensuring complete ejection of the cut out part.

Face cut label – Any pressure-sensitive label in which the face material has been cut to the liner. A die-cut label product from the waste matrix around the labels has not been removed.

Flex – The deflection of rollers or cylinders in a printing press. Also describes the bending qualities or characteristics of any material, including printing plates. See deflection.

Flexible dies – A thin, flexible steel die-cutting 'foil' or plate for use on magnetic cylinders, magnetic base or other special die-cutting systems. Flexible cutting dies are etched from specially formulated steel ranging from 0.5 – 1.5 mm in thickness. Flexible dies are lower in cost than solid dies and the economics of use become more attractive as the complexity of the label shape increases. The life is much the same as for a solid die, providing cleanliness, setting, anvil condition, adhesive and label design are properly controlled. Flexible dies are available today with a wide range of surface treatments and provide good results with a wide variety of materials.

Friction – is the force resisting the relative motion of solid surfaces and is usually proportional to the amount of contact force which presses the surfaces together as well as the roughness or the texture of the surfaces.

Hardness – Hardness is a very important property of tool steel that is developed during the heat treating process. It is not one of the inherent properties of tool steel. Hardness is developed through the addition of carbon.

Heat resistance – The ability of tooling steel to resist softening when exposed to heat while in operation. This is a very important property in high-speed steel tools. Excessive heat in a turning or milling operation can lead to the softening of the tool. This softening allows the tool to dull or chip causing premature failure.

Hole punching (pinfeed) is normally used for making EDP holes but other shapes are also possible. The holes are cut by using either a male / female system or using a shaft with movable EDP rings. The latter is used in anvil position to cut onto the face material and the waste is being removed by the waste matrix when stripping the face.

Hot foil unit – The part of a label press that contains the equipment to hot foil self-adhesive material.

Induction hardening – A form of heat treatment in which a metal part is heated by induction heating and then quenched. The quenched metal undergoes a martensitic transformation, increasing the hardness and brittleness of the part.

Journal – The shaft on the operator and gear side of a cylinder that is supported by a bearing

Key – In mechanical engineering, a key is a machine element used to connect a rotating machine element to a shaft. Through this connection the key prevents relative rotation between the two parts and allows torque to be transmitted through. For a key to function the shaft and rotating machine element must have a keyway, which is a slot or pocket for the key to fit in. The whole system is called a keyed point. A keyed point still allows relative axial movement between the parts.

Kiss cut – A die-cutting operation in which self-adhesive face material is cut through to the release liner backing, but the liner itself is not cut.

Laser hardening – Used as an alternative to the more common surface treatments, some suppliers of flexible dies also offer laser hardened dies. Laser hardening is not offering any reduction in the friction coefficient, but does offer a partial increase in hardness at the tip of the cutting edge. The increase in hardness depends on the carbon content in the steel and not on the energy put in by the laser.

Magnetic cylinder – A stainless steel-based cylinder having a series of permanent magnets glued around its periphery and used in die-cutting to hold flexible dies in place. Magnetic cylinders will fit on to any press that takes rotary dies, and there are no size limitations outside of those relating to the press dimensions. They are installed in exactly the same way that conventional rotary dies are installed.

Matrix waste – The skeleton of face material and adhesive waste surrounding self-adhesive labels after die-cutting. The matrix waste is normally removed on the press converting line by stripping it from the web of die-cut labels and re-winding it on matrix re-wind roller. Waste may also be removed by shredding into small pieces and/or by extraction systems that convey the waste away for bagging or baling. Also known as the waste skeleton.

Micro perforation – Very small perforations or minute pinholes in paper that enable a section or part of the paper to be easily separated. Micro perforation leaves a smooth edge without the normal more jagged edge found with standard methods of perforation.

Operating temperature – An operating temperature is the temperature at which a label press operates. The press will operate effectively within a specified temperature range which varies based on the device function and application context, and ranges from the minimum operating temperature to the maximum operating temperature (or peak operating temperature.

Oxidation – result of a reaction between tool face and oxygen exposure. In order to prevent tooling to rust (oxidation) please clean and oil the tool carefully before storage. Oxidation can have a detrimental effect on the cutting results of both rotary dies and well as flexible dies.

Perforation – A line or row of cuts or tiny holes that enable a paper or web of labels to be folded, torn off or separated easily. Perforation may be horizontal or vertical with standard or micro perforations (cut versus tie).

Razor slitting – Mostly used for slitting thin plastic films – these type of systems are very simple, quickly and easy to set. Although razor blades are a low cost product, they need to be frequently changed to ensure a good quality slit edge.

Rotary Pressure (RP) cutting – a shear type of cut made by passing two precisely machined cutters by or through one another. The material is actually squeezed or compressed to the point of bursting without the two parts of the tool ever touching. The two cylinders that make up the tool set both rotate at exactly the speed in order to create a perfect match. Not very common in the narrow web industry. Only for cut through applications and (very) long runs.

Rule cutters – Rule cutter are used for flat die-cutting in presses where the web is momentarily halted during the actual cutting process. Consequently, the output speed is somewhat slower than when rotary tooling is used.

In its most simple form, the flat ruled cutter may be produced by working to dimensions, obviating the need for a drawing when the shape is a simple square, rectangle or circle. For more complex cutting shapes a key line can be drawn by hand or created by some form of electronic origination at the same time as the label design is created.

Shear cut – Used to describe the cutting of a continuous web of stock using an action similar to the action of scissors.

Shear slitting – Shear is the most versatile slitting method. Shear slitting can be used for a wide variety of papers, films, laminates and foils. Shear slitting variations can be 'tangential/kiss' or 'wrap' shear which involves a loaded male blade against a female ring, creating a scissoring effect to slit the web. Shear slitting typically provides the highest quality edge quality.

Sheeting – Converting rolls of printed or unprinted label-stock into individual sheets on a roll-fed press.

Slitter/re-winder – The use of slitter/re-winder technology is one of the principal methods of off-line converting rolls of self-adhesive labels, tapes and flexible packaging film into the correct width rolls ready for shipping to customers. The operation may also be combined with web inspection as part of the final stage in a converter's quality control program.

Slitting – The action of cutting rolls of label stock to specified widths on a slitting machine. Slitting is undertaken using either stationery or rotary knives or blades in a machine with roll unwind and rewind devices, tension control and web tracking. Slitting of master rolls to narrow-web roll widths is normally undertaken by the label stock manufacturer, whose machinery has lubricated slitting knife cleaning pads on each rotary blade to stop adhesive build up on these slitting tools. These slitting machines also have a special splicing table, before the slitting station, where a diagonal butt splice can be made after removal of pre-marked sub-standard material, plus paper mill and coating machine joins.

Slitting wheels/knives – Dividing a web of label stock in the lengthwise direction by rotating slitting knives so as to produce two or more narrower webs. Slitting may be carried out with shear or crush cutting. Shear cutting produces a good quality edge a high line speeds with little dust; crush cutting can be economical if speed and edge quality are less critical.

Standard air eject dies (through shaft bore) – Air eject cylinder which uses compressed air to blow out of the holes drilled in the various cavities. For these types of cylinders there is little/no control over the waste ejection.

Stepped anvil – An anvil with the bearer area either higher or lower than the main body. Stepped up – bearer is lower than the body, creating a deeper cut. Stepped down – body is lower than the bearer, creating a more shallow cut.

Stripping – Also called waste stripping. Removal of the face material and adhesive (the matrix waste) from around the die-cut label by taking it around a roller, or over a metal bar, prior to being re-wound.

See also Matrix waste.

Thru cut/Through cut/Cut through – The action of die-cutting through all the layers in a pressure-sensitive laminate. This may take place in just one part of the cutter profile, or it may involve the complete profile.

Tolerance – the permissible deviation from a specified value. This applies e.g. to flexible die height, calliper of a backing.

Tool(ing) steel – Tool steel refers to a variety of carbon and alloy steels and that are particularly well-suited to be made into tools. Their suitability comes from their distinctive hardness, resistance to abrasion their ability to hold a cutting edge, and/or their resistance to deformation at elevated temperatures (red-hardness). Tool steel is generally used in a heat-treated state.

Trim – Trim is used to describe the normal edge waste that is removed from a master roll of label stock. On the printing and converting line, trim describes an action on the press, i.e. edge trim, waste trim.

Vacuum die – Rotary dies with removable inserts. Instead of blowing out the waste, the cut out particles are sucked away through the journals by means of vacuum.

Wear resistance – The ability of tooling steel to resist erosion. Wear resistance in tool steel is achieved by the presences of carbides. Chromium, molybdenum, tungsten and vanadium are the four carbide-forming elements commonly found in tool steels.

Web tension control – the amount of pull or tension applied to the web as it passed through the press. Poor tension control will result in registration problems in the printing, embellishing and converting processes.

Index

www.ingramcontent.com/pod-product-compliance
Lightning Source LLC
Chambersburg PA
CBHW041720210326
41598CB00007B/715